# Instrument Flying

## Other McGraw-Hill Aviation Titles

*Lewis Bjork*
PILOTING FOR MAXIMUM PERFORMANCE

*Robert N. Buck*
WEATHER FLYING, 4TH EDITION

*Jerry A. Eichenberger*
YOUR PILOT'S LICENSE

*Tony Kern*
REDEFINING AIRMANSHIP

*Shari Stamford Krause*
AVOIDING MID-AIR COLLISIONS

*Wolfgang Lanewiesche*
STICK AND RUDDER: AN EXPLANATION OF THE ART OF FLYING

*Michael C. Love*
BETTER TAKEOFFS AND LANDINGS

*John F. Welch, Editor*
VAN SICKLE'S MODERN AIRMANSHIP, 7TH EDITION

# Instrument Flying

## Richard L. Taylor

**Fourth Edition**

**McGraw-Hill**

New York   San Francisco   Washington, D.C.   Auckland   Bogotá
Caracas   Lisbon   London   Madrid   Mexico City   Milan
Montreal   New Delhi   San Juan   Singapore
Sydney   Tokyo   Toronto

**Library of Congress Cataloging-in-Publication Data**

Taylor, Richard L., 1933–
    Instrument flying / Richard L. Taylor.—4th ed.
        p.    cm.
    Includes index.
    ISBN 0-07-063345-2 (alk. paper)
    1. Instrument flying.  I. Title.
TL711.B6T36  1997
629.132'5214—dc21                  97-19404
                                    CIP

# McGraw-Hill

*A Division of The McGraw·Hill Companies*

1 2 3 4 5 6 7 8 9 0  FGR/FGR  9 0 2 1 0 9 8 7

ISBN 0-07-063345-2

*The sponsoring editor for this book was Shelley Ingram Carr, the editing supervisor was Ruth W. Mannino, and the production supervisor was Sherri Souffrance. It was set in Garamond by McGraw-Hill's desktop publishing department in Hightstown, N.J.*

*Printed and bound by Quebecor/Fairfield.*

McGraw-Hill books are available at special quantity discounts to use as premiums and sales promotions, or for use in corporate training programs. For more information, please write to the Director of Special Sales, McGraw-Hill, 11 West 19th Street, New York, NY 10011. Or contact your local bookstore.

This book is printed on recycled, acid-free paper containing a minimum of 50% recycled, de-inked fiber.

*The Fourth Edition of* Instrument Flying *is dedicated to my second grandson, Cameron Charles Taylor, who will no doubt observe—and I hope experience—advances in instrument flying that I can only imagine.*

# Contents

# Contents

# Contents

# Foreword

*Robert N. Buck*

This book is a valuable link between theory and what instrument flying is really like. It's a book to read all the way through and also one to have handy for those free moments to pick up, open anywhere, and gain a useful piece of information.

*Instrument Flying* is not restricted to any level of pilot experience. It will interest the person who is just beginning to think about getting an instrument rating, and it will be valuable to the experienced pilot as well.

It is difficult to dig out all the information about instrument flying. I've tried by reading stuffy technical books, the stiff FAA publications, and formal study courses designed to get one through the FAA exams. And even after all this digging, there is still a long way to go to discover what it's really like—to know what's behind the scenes. Mostly this comes with experience.

*Instrument Flying* cuts this process short because it tells, as it teaches, all about the inside and what's behind the formal stuff. In doing this, the book gives experience as it tells. I wish it had been around years ago; it would have made things a lot easier.

Richard Taylor has an excellent background in flying and in education. He has had more than 40 years' experience as a military and then as a commercial pilot. Along with his practical background, he is an associate professor emeritus in the Department of Aviation at the Ohio State University. The book reflects this fortunate combination as he takes us from attitude flying, which is the solid basis for flying on instruments, to the sophisticated techniques of the high-altitude airways.

The complex problems of absorbing a clearance, using the radio, and staying within the law are presented in a straightforward, engaging style that will help lift pilots from a timid, unsure position to one where they can operate like pros. The chapters on holding patterns and instrument approaches, plus all the aspects of a flight from A to B on instruments, are extremely valuable for learning and attaining proficiency through practice.

This book is an important addition to every pilot's library. But it is not a book to leave on the shelf gathering dust; rather it will be read many times and, I'm certain, consulted time and time again to refresh your memory and settle many a friendly argument.

# Preface

"Things are always and never the same," in the words of an anonymous philosopher who was apparently an accurate observer of the human experience. Whether the reference was to the climate or the moods of humanity is immaterial, for the more things change the more they exhibit characteristics that have always existed.

This philosophy is certainly applicable to the rather specific subject matter of this book. Even though we have seen remarkable advances in aerodynamic, propulsive, and avionics technology, the very basics of flight still apply to instrument flying. After all, we go through essentially the same motions to get an airplane from here to there in the clouds as the Wright brothers did in 1903 when they first flew.

So what's new in this fourth edition of *Instrument Flying?* I've retained all the fundamental information that continues to apply, while adding a number of regulatory and procedural changes that affect day-to-day IFR operations in the national airspace system. As in the previous editions, I have tried to make the book as practical as possible; that is, to make you aware of the way things work in the real world of instrument flying. You'll find shortcuts that aren't in "the book," and other procedures that should improve your efficiency and make the IFR experience more enjoyable.

Virtually every nonfiction aviation book promotes safety of flight— it may be a stated objective or a subtle undertone—and *Instrument Flying* is no exception. I believe that one of the best ways to fly safely is to study and profit from the mistakes of others. With that in mind, the fourth edition includes more vignettes of actual incidents that are intended to help you stay out of trouble by recognizing and avoiding situations that may show up when you're flying IFR. Please understand that my comments are not intended to pass judgment on the quality or performance of any individuals, groups, aircraft, or equipment; the sole purpose is to illustrate the consequences of certain decisions and procedures in the conduct of IFR operations.

If you're a beginning IFR pilot, I hope this book will help you to obtain the rating; for those of you already involved in instrument operations, I hope that *Instrument Flying* will be helpful in improving your IFR skills and your ability to operate more efficiently in the airspace system.

Richard L. Taylor
August 1997

# Instrument Flying

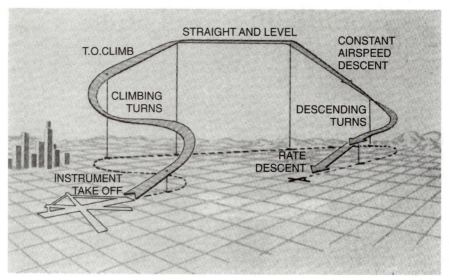

*Any instrument flight, regardless of how long or how complex, is simply a series of connected basic instrument flight maneuvers.*

# 1

# The Complete Instrument Pilot

Put a foot-wide steel beam flat on the ground and walk across it; a "no sweat" situation for anyone with normal balance and eyesight. Now put that steel beam between two buildings 10 stories above the street, and anyone less than an experienced steelworker or a professional highwire performer would panic at the prospect of negotiating the same narrow path that presented no problem at ground level. The difference? Knowledge, experience, and practice.

Much the same reasoning applies to instrument flying; you know that you can handle your airplane when you can see the ground, but getting from here to there in the clouds is something else. And no matter how well you fly the machine, there's always Air Traffic Control (ATC), giving you sometimes confusing instructions, asking you to maintain cruise airspeed on an approach, clearing you to an altitude you don't want, and on and on and on. But basic principles always apply, and if you are well grounded in the "nuts and bolts" of instrument flying, there's no reason why you can't adjust to changes and involved procedures *if you know what to expect, and how to handle yourself in the IFR system*. There's nothing heartstopping about flying an airplane on instruments, nor does it take a superman to do the job well. Good instrument training plus a thorough understanding of the *total* system can make it just as easy as walking that steel beam when it's flat on the ground. When you know what you're doing, you can walk it confidently and safely 10 stories high.

More than 40 years of military and civilian flying in many parts of the world, and in all kinds of weather, have convinced me that once past the fundamentals, safe and efficient instrument flight is a happy combination of knowing yourself, your airplane, and the system. When you accept a clearance from ATC to fly in the nation's airspace, you are considered just as qualified, just as capable as the airplane captain who flies every day. This is not to imply that controllers are so unrealistic that they expect your approaches to be as exquisitely precise as the "pros," but all IFR pilots *are* expected to conform to the same rules and procedures because they become part of a *system*—a combination of parts into a whole, an orderly arrangement. The principal parts of this system—controllers and pilots—must work together if the system is to accomplish its goal of safe separation and efficient management of the thousands of instrument flights conducted every day.

The key to the whole process is knowledge; it has been proven that the average pilot can be taught to fly an airplane by referring only to the attitude instruments (indeed, it's a required part of the practical examination for the private pilot certificate). But throw in the complications of electronic navigation and rapid-fire communication, and average pilots can come unglued...unless they know what the system is all about. "You learn something every day" is a bromide perhaps more meaningful in instrument flying than in any other endeavor, for regulations, procedures, and techniques change almost daily. The aviation community suffers plenty of "understanding gaps" which derogate pilot performance, efficiency, and sometimes compromise safety.

Years ago, I was involved in programs designed to refresh the knowledge of instrument-rated pilots and to prepare candidates for written examinations relative to IFR and ATP operations. That experience convinced me that the surface of pilot education has barely been scratched and that there is a profound need to continue the training *after* pilots obtain their instrument ratings. *Instrument Flying* is intended to help put a real gouge in that surface; it has been developed with the philosophy that knowledge, added to practice and experience, will pave the way to safer and more efficient instrument flight.

This book can be used to its greatest advantage as a source of information for understanding the *total* system in which we fly IFR. In addition to some basic techniques of aircraft control, you will find detailed explanations of every phase of instrument flight, from Airways to Zulu time. To help you increase the efficiency and utility of your airplane, *Instrument Flying* contains techniques and procedures for practical, legal methods of cutting down the elapsed time between point A and point B; isn't that the real reason for using an airplane in the first place?

The chapters are strung on a common thread of increased usefulness, of maximizing the dollars you spend to transport people and things through the air. Chapters dealing with attitude instrument flying are not there to teach you how to fly instruments, for that is your flight instructor's job; however, you can and should use this material as a guide for practicing and polishing your flying skills. "Good" instrument pilots are the ones who fly their airplanes without conscious effort, conserving the major portion of their thought processes for navigation, communication, and staying at least one step ahead of the airplane at all times.

## ✛ An Instrument Book with No Chapter on Weather?

When you can't beat 'em, join 'em! Robert N. Buck, now retired from his position as senior captain with Trans World Airlines, has put together the ultimate interpretation of aviation meteorology for the instrument pilot, so I defer to his work in this vital area, which has more to do with IFR operations than anything else. If you do not own a copy of Buck's *Weather Flying* (4th ed.; McGraw-Hill), I strongly recommend that you add it to your aviation library. Seldom do we have the opportunity to share in, and profit from, the vast experience of one so eminent in his field.

## ✛ Be a *Legal* Eagle

When the sands at Kitty Hawk were brushed by the skids of the Wright brothers' airplane, when theirs was the only powered flying machine in the entire country, there was no need for rules of the air— the laws of gravity kept Wilbur and Orville busy enough. But soon there were two airplanes, and then more and more and more; just like the increased traffic on rivers and roads, a burgeoning aircraft population eventually had to come under regulation and control. Safety of flight has been uppermost in the minds of rule-writers from the very beginning (check the aviation regulations—almost all of them are intended to make flying safer), with a secondary purpose of helping to establish who's at fault when an accident *does* occur.

The regulations are therefore either restrictive or mandatory in nature, letting us know those things we may *not* do or the things we *must* do. Although the intent remains steadfast, the language and scope of aviation regulations change constantly, as the nature of flying itself changes. It would be a fool's task to list all the rules that apply to instrument flying because some of them will change before the ink on these pages is dry. The regulations (and revisions thereto) are available to all, and so smart instrument flyers (or smart *any*-kind-of-flyers, for that matter) will subscribe to those that apply to their operations, and moreover will keep their books right up to date. For noncommercial pilots (VFR *and* IFR types), Part 91 of the Federal Aviation Regulations is a bare minimum, and the *Aeronautical Information Manual* (*AIM*) will help you stay current in regard to changes in procedure and technique. Jeppesen's "J-AID" combines the several parts of the regulations, all the information contained in Part I of the *AIM,* and a great deal of additional information you may find worthwhile—IFR and VFR.

"But," you protest, "I haven't time to spend going through regs and manuals—I fly good instruments, do what ATC says, keep my medical current and my nose clean." That's fine as far as it goes, but do you have the time or the resources to contend with a judgment against you as the result of a violation? As in any court action, ignorance of the law is never an excuse, especially when one of the very first parts of our aviation rules says that the pilot in command will, in effect, be familiar with *everything* that may affect the operation of an aircraft before a flight begins. It takes only one small slip on your part to render yourself defenseless—there is normally no way to be involved in an accident or incident with an airplane and not be in violation of some part of the aviation regulations. Unfortunately, everyone else in the world figures that a person who can afford to own or operate an airplane can also afford a huge settlement. And after you lose all your money in a court action, the government may step in and relieve you of your flying privileges—sometimes permanently.

There's a practical side to knowing the regulations, too. With the ever-increasing variety of instrument approaches and other system options that we can put to use, less-than-current IFR pilots will sooner or later come up against a situation that could be avoided by knowing just what they may or may not do. Current knowledge and complete understanding make the difference between pilots who stumble through the airspace confused and bewildered and those who make things happen for their benefit, efficiency, and safety.

The point of all this is that you *must* know the rules of the flying business, you *must* stay abreast of changes as they are effected, and it's *more* important when you are IFR. VFR-only flyers can get by with less regulation reading because they can always rely on the see-and-avoid rule. But when you are accepted into the IFR system, you must play in the same key as everyone else; from Ercoupes to airliners, instrument pilots must operate on a common base of regulation and control. Now that you sometimes can't see where you're going, it's comforting to know that all the other pilots up there in the clouds with you are flying by the same set of guidelines.

And what about the controllers' rules? While you can't be held responsible for knowing all the hoops they must jump through, the chances are excellent that you'll become a more understanding and system-oriented pilot once you've observed the conditions in an Air Route Traffic Control Center or Approach Control facililty. They sponsor periodic public-training sessions (usually known as "Operation Raincheck") which include an extensive tour of the facility and a series of classes aimed at developing better cooperation between controllers and pilots. The local Flight Standards Office or any handy FAA facility should have a schedule of such programs.

## ✛ Your Choice of Charts

There are two popular sources of IFR charts: one is the federal government; the other is Jeppesen Company, a commercial supplier. Both services include a wide range of publications, from SIDs to STARs and everything in between. You can purchase charts for the whole world, or only for the area in which you fly; it's entirely up to you and your pocket. Which service is best? That's an impossible question because pilot preferences vary so widely. Get a sample of each, and try them—that's the only way you'll be able to decide which is best for you.

No matter what your choice, there is one common denominator: IFR charts cannot fulfill their ultimate purpose unless you know what every mark thereon means. New symbols are added, airway courses are adjusted, and radio frequencies are changed, so it becomes nothing short of mandatory that you keep yourself up to date with current charts and full knowledge of those charts. One small, seemingly inconsequential bit of information might just save your neck some day.

Realizing the same out-of-date-before-the-ink-is-dry problem that exists with regulations, *Instrument Flying* does not include a chapter on IFR charts. On occasion, it is necessary to illustrate a particular point with symbols or numbers that are expected to remain in use indefinitely—but you must realize that you are expected to know the sometimes restrictive, sometimes permissive, sometimes directive nature of all the chart symbols and markings. Get *very* familiar with the legend pages supplied with your charts. The Jeppesen Company treats this problem thoroughly in the respective sections of its IFR chart publications. If you use the government charts, you would do well to get a copy of "Civil Use of U.S. Government Instrument Approach Procedure Charts" (an FAA Advisory Circular) and learn it inside out.

Whenever you receive revisions to your charts, *insert them right now,* and do it yourself—don't trust anyone else. Allowing 3 or 4 weeks' revision notices and new charts to pile up on your desk has two bad features: First, it's very frustrating to replace the same chart several times at one sitting; second, anytime you're flying IFR with outdated charts, you're asking for trouble. And when you insert those revisions, pay attention to the changes that caused the revision; you'll learn a great deal about the system by taking note of the new procedures.

## ✝ Advisory Circulars

Periodically, the FAA issues information of a nonregulatory nature in the form of Advisory Circulars. A subscription to the appropriate parts of this information system will help keep you up to date on the latest official thinking in several areas of interest. For starters, you might consider these four (each subject area is followed by the number that identifies it in the FAA system): Aircraft (20), Airmen (60), Airspace (70), Air Traffic Control and General Operations (90). Advisory Circulars can be ordered from the U.S. Government Printing Office, Washington, D.C. 20402, or any one of the government bookstores around the nation—and saving the best for last, most of them are *free!*

## ✝ That Still, Small Voice

There's really not much that can get you into serious trouble on an instrument flight except weather and the consequences that develop around it. If your airplane is properly maintained, checked, and operated, chances are excellent that it will get you where you want to go. But sometimes, when "get-home-itis" is coming on strong, and your IFR capability is bolstering a "nothing-can-stop-me" frame of mind, listen to that still, small voice of reason that reminds you to back off and take another look at the situation. The temptation to press on in the face of bad weather has led many a pilot down the garden path; if you've worked out a Plan B to take care of nonflyable weather, you can make the decision well ahead of time, still get your business accomplished, and come back to fly another day.

There will be times when you cancel a flight only to have the weather gods make a fool of you with a beautiful day—those are the breaks of the game. But when you've left yourself no way out and make a forced decision to take off into marginal weather that turns out *worse* than expected, you're in the wringer. If you're lucky, you may get by with nothing more than a harrowing experience and a monumental resolve never to do that again. It's a lot more comfortable to be on the ground wishing you were in the air than to be up there wishing you were on the ground! Professional pilots earn their pay, and amateurs earn the respect of their passengers, on those occasions when they overcome pride with good sense and say, "The weather looks worse than I want to tackle today—we're not going by air."

# ✢ Have at It!

There is no recommended order for reading *Instrument Flying*; dig right in wherever you feel your knowledge is a bit rusty or when you come across a situation you don't understand thoroughly. Each section is functionally complete in itself and does not depend on previous study of some other part of the book; other chapters may be consulted for more detail, and the Glossary in Chap. 2 is available for definition of terms.

Now I have your clearance; are you ready to copy? You are cleared from here to the end of this book, via 21 chapters packed with instrument flying information. Climb to and maintain a higher level of efficiency and safety; expect further clearance as you work through these pages and become a *complete* instrument pilot. Have a good flight!

# 2

# The Language of Instrument Flying

Passengers who listen to the radio chatter during an instrument flight probably feel as if they have been exposed to a replay of the Tower of Babel scene. The working language, the jargon, of instrument flying does get a bit confusing at times; but then, the at-work conversation of *any* specialized occupational group will sound like glossolalia to the uneducated.

Terms, definitions, and contracted phrases can pile up into a meaningless mess of initialese and confusion when you're about the business of conducting a flight under instrument conditions. When a timesaving communications shortcut comes your way and you're not sure just what you are expected to do, or a term crops up which is completely foreign, by all means *ask*; the worst thing to do is *assume* that you know what something means—you can get into a lot of trouble doing that.

There are two ways to become fluent in the language of instrument flying; one is to fly IFR at every opportunity and gradually pick up the jargon, and the other is to fly IFR at every opportunity and gradually pick up the jargon. But *first* know what you and they (ATC) are talking about by studying the following glossary. It's a collection of frequently used phrases, definitions, abbreviations, and acronyms that will help you understand what ATC is saying to you and improve the propriety of your electronic conversations. There are some terms which you will probably never use, but which are included to accommodate the increasing sophistication of instrument flying. Hold on to your spoon—the alphabet soup promises to get thicker and thicker. (Refer to the *AIM* Part I or the "J-AID" for a more detailed listing of aeronautical terms.)

# ✝ Glossary

**abeam** A position directly off either wingtip—a relative bearing of 90 or 270 degrees.

**ADF (Automatic Direction Finder)** Refers to the low/medium frequency radio receiver in the airplane. The ADF indicator gives the pilot a readout of the bearing from airplane to station.

**affirmative** Better than "yes" because it is more easily understood.

**aircraft classes** Used by ATC for purposes of wake turbulence separation, all aircraft are classified according to maximum takeoff weight as follows: heavy—more than 255,000 pounds; large—41,000 to 255,000 pounds; small—41,000 pounds or less.

**AIRMET** An advisory of weather conditions considered potentially hazardous to light aircraft; usually high winds, low ceilings, low visibilities.

**airspeed** Velocity of an aerial machine; may be stated in a number of ways (always knots): *Indicated*—The number to which the needle points. When a controller asks, "What is your airspeed?" respond with the number of *knots* at the end of the pointer. *Calibrated*—Pointer indication corrected for installation and instrument error. All limiting and performance speeds are quoted in terms of calibrated airspeed. *True*—Calibrated airspeed corrected for pressure and temperature; the actual speed of the airplane relative to undisturbed air. This is the speed used for IFR filing purposes. *Basic Rule When in Flight*—"Maintain thy airspeed, lest the earth arise and smite thee."

**airway** Designated air route between points (usually VOR transmitters) on the earth's surface.

**ALS (Approach Lighting System)** An arrangement of lights designed to provide visual guidance to a pilot breaking out of the clouds on an instrument approach. There are a number of acceptable displays; the most striking feature of a typical installation is the "ball of fire" effect from the long line of high-powered sequenced flashers cascading toward the runway.

**alternate airport** A place to go if weather at the airport of intended landing goes sour. Must be part of an IFR flight plan under certain conditions.

**altimeter setting** The barometric pressure reading of a given point on the surface, reduced to sea level. Used to adjust altimeters for variations in atmospheric pressure and to standardize all altimeters for high-altitude flight (all altimeters are set to 29.92 inches above 18,000 feet MSL).

**altitude** Available in several models, including: *Pressure Altitude*—Read on the altimeter when the altimeter setting is 29.92 inches; everyone operating above 18,000 feet uses pressure altitudes. *See also* FL (Flight Level). *Density Altitude*—Pressure altitude corrected for temperature; this is the altitude at which the airplane thinks it's flying and the altitude upon which all performance figures are based. *Indicated Altitude*—What you see on the altimeter when the current setting is placed in the window; all assigned altitudes below 18,000 feet are indicated altitudes. *Absolute Altitude*—Your actual height above the terrain. Used to determine decision height for Category II and III approaches, and sometimes used for over-water navigation. *Radio Altitude* (often incorrectly called "radar altitude")—Same as absolute altitude; a small transceiver is used to measure height above the surface. *True Altitude*—Your actual height above sea level.

**approach category** Grouping of aircraft according to a computed speed and maximum landing weight to determine adequate airspace for maneuvering during a circling instrument approach procedure.

**approach speed** A computed number used to determine the category (A, B, C, D) to be used for instrument approaches. It is the calibrated power-off stall speed of the aircraft at maximum landing weight in the landing configuration, multiplied by 1.3.

**arc** A circle of constant radius around a VORTAC station. When a DME arc approach is specified, you will fly around the station at a fixed distance until intercepting an approach radial which will lead you to the airport.

**area navigation** *See* RNAV.

**ARO (Airport Reservation Office)** The FAA facility that manages requests for flight operations (IFR and VFR) to and from designated high-density airports (JFK, LGA, ORD, DCA).

**ASR (Airport Surveillance Radar)** Relatively short-range radar equipment used primarily for approach control in the terminal area. May also be used as an approach aid to vector aircraft to within 1 mile of a runway.

**ATC (Air Traffic Control)** Any facility engaged in the direction and control of aircraft in controlled airspace; this includes Clearance Delivery, Ground Control, Tower, Approach Control, Departure Control, Air Route Traffic Control Center, and Flight Service Stations.

**ATIS (Automatic Terminal Information Service)** A continuous broadcast of data pertinent to a specific terminal; includes weather, altimeter setting, approaches, departures, runways in use, other instructions, and general information.

**AWOS (Automated Weather Observation System)** Senses and broadcasts local weather conditions on a published VHF frequency. Depending on system type, an AWOS broadcast may include altimeter setting, wind direction and velocity, temperature, dewpoint, density altitude, visibility, and cloud/ceiling data.

**back course** The "other side" of an ILS Localizer course: the electronic extension of the runway centerline, proceeding in the opposite direction from the front course. Most back courses provide an additional nonprecision approach for the airport, and some have glide slopes.

**back course marker** A range indicator similar to the outer marker, but located on the back course. Provides distance-from-the-runway information and is usually designated the FAF for that approach procedure.

**bearing** The relative position of one object to another, stated in degrees. For instrument navigation, a bearing means the direction *toward* a nondirectional beacon. All ADF approach charts show bearings TO the station; VOR instructions are always in terms of *radials*—bearings AWAY from the station.

**CAVU (Ceiling And Visibility Unlimited)** Nothing more than scattered clouds; visibility more than 10 miles. Not an official Weather Service term.

**CDI (Course Deviation Indicator)** In the vernacular of the everyday pilot, it's known as the left-right needle on the VOR display.

**ceiling** The lowest broken cloud layer not reported or forecast as "thin."

**cell** In conjunction with a radar advisory or report, implies a strong echo and *usually* indicates a thunderstorm. To stay on the safe side, always consider a reported cell a thunderstorm and request vectors around it.

**Center** The ATC facility responsible for the enroute phase of IFR operations; the full name is Air Route Traffic Control Center.

**circling approach** Any instrument approach in which the runway to be used for landing is aligned more than 30 degrees from the final approach course, or when a normal rate of descent from the minimum IFR altitude to the runway cannot be accomplished. Clearance for such an approach will always be specific, for example, "cleared for the runway 10R ILS approach, *circle to land* runway 15." A circling approach is *always* a nonprecision approach.

**Class A airspace** All U.S. airspace from 18,000 feet MSL up to and including FL600, in which all flight operations must be conducted in accordance with instrument flight rules.

**Class B airspace** In general, airspace from the surface to 10,000 feet MSL above the nation's 12 busiest airports. An ATC clearance (IFR or VFR) is required to enter, transit, or depart this airspace.

**Class C airspace** In general, airspace from the surface to 4000 feet above airports with operating control towers and radar approach control facilities. Two-way radio communication must be established prior to operating in Class C airspace. Therefore, it is of little interest to IFR pilots.

**Class D airspace** In general, airspace from the surface to 2500 feet above airports with operating control towers but no radar service. Two-way radio communication must be established prior to entering. Again, of little interest to pilots operating on IFR clearances.

**Class E airspace** All controlled airspace not designated as Class A, B, C, or D. May begin at the surface for some airports with published instrument approach procedures, or may begin at either 700 or 1200 feet AGL as transition areas for IFR arrivals and departures. All federal airways are Class E airspace.

**Class F airspace** Does not exist in the United States.

**Class G airspace** All uncontrolled airspace. IFR flights are permitted in Class G airspace, but no ATC separation services are available.

**clearance** An authorization from ATC to proceed into or through controlled airspace. A clearance provides, changes, or amends your limits in any one, or all three, dimensions of flight: altitude, route, and point to which you are cleared.

**clearance limit** An electronic fix (intersection, NDB, VOR, DME fix) beyond which you may not proceed in IFR conditions without further clearance. Never accept a clearance limit unless you receive an expect-further-clearance time.

**clearance-void time** The latest legal takeoff time when you have been issued an IFR clearance at a nontowered airfield.

**cleared as filed** A communications simplifier that means you are cleared to the destination and via the route you requested in your flight plan. Does *not* include an altitude, which must be assigned separately.

**cleared direct** An ATC instruction that means proceed from your present position in a *straight line* to the appropriate fix.

**cleared for straight-in approach** Proceed direct to the final approach fix and complete the approach without executing a procedure turn.

**cleared for the approach** Proceed from your present position direct to the appropriate radio facility and execute the approach as published.

**compass locator** A nondirectional low-frequency radio beacon colocated with the outer marker, providing a signal which you can use to navigate with ADF to the OM. Also called an "outer locator," "outer compass locator," or just plain "locator." Same as LOM.

**composite flight plan** A combination flight plan that contains IFR and VFR segments. Often used to expedite flight operations when VFR conditions exist at departure or destination, or vice versa.

**contact approach** A shortcut to a published instrument approach procedure, in which a pilot may be cleared to the destination airport using ground references.

**course** A line drawn on a chart between two points and, when given a direction, is referenced to either true or magnetic north. All courses on IFR charts are *magnetic*.

**cruise clearance** Always implies a clearance for an approach at the destination airport, as well as important altitude and airspace limitations.

**CTAF (Common Traffic Advisory Frequency)** The published communications frequency to be used by all pilots (IFR and VFR) operating at an airport with no ATC communications. CTAF is usually a Unicom frequency at nontowered airports and the local-control frequency when the tower is closed.

**DF (Direction Finding)** A disoriented IFR pilot's last resort. Most Flight Service Stations have the capability of electronically determining your bearing and supplying headings to fly to the airport. An approach aid under emergency conditions, DF is one way of becoming "unlost."

**DH (Decision Height)** A point on the guide slope determined by the altimeter reading. Upon reaching the DH (an altitude published in feet above sea level), a decision must be made to either continue to a landing or execute the missed approach procedure. Decision heights are associated *only* with precision approaches.

**DME (Distance-Measuring Equipment)** An airborne navigational aid which interrogates a VORTAC or VOR/DME station and, depending on the sophistication of the black box in the airplane, can provide distance, groundspeed, and time-to-station. Frequently, DME is used to identify intersections and confirm locations.

**EFC (Expect Further Clearance time)** Always issued to holding or clearance-limited IFR flights to provide time for controllers to clear airspace ahead in the event of communications failure.

**ETA** Estimated Time of Arrival.

**ETD** Estimated Time of Departure.

**ETE** Estimated Time Enroute.

**FAF (Final Approach Fix)** The last radio- or radar-determined position before you begin descent to the minimum altitude for an instrument approach. A report should be made when passing over the FAF inbound during an approach.

**fan marker** A highly directional radio transmitter used to indicate distance from the runway on an approach. The radiation pattern seen from above would look like a football with pointed ends. Outer markers, middle markers, and inner markers are fan markers. Usually shortened to "marker."

**final approach** That segment of an instrument approach procedure that leads from the final approach fix to the missed approach point.

**fix** A definite geographical position that may be determined by the intersection of bearings from radio navigation stations, by radar observations, by use of VOR and DME, or any RNAV system.

**FL (Flight Level)** Term used to indicate pressure altitudes to be flown in the high-altitude route system (18,000 feet and above).

**Flight Watch** The code name for contacting specially trained Flight Service Specialists for weather information. The universal frequency to be used below 18,000 feet is 122.0.

**FSS** Flight Service Station.

**glide slope** An electronic signal that provides vertical guidance during a precision approach.

**GMT (Greenwich Mean Time)** *See* Universal Coordinated Time.

**go-around** For IFR pilots, same as "missed approach."

**GPS (Global Positioning System)** A satellite-based navigation system that provides highly accurate position, velocity, and time information.

**"hard" altitude** An altitude *assigned* by ATC; the actual altitude is always preceded by "climb to and maintain" or "descend to and maintain."

**heading** The direction in which an aircraft is pointed, related either to true north or magnetic north. In domestic IFR operations, true headings are never used—everything is magnetic.

**heavy** *See* aircraft classes.

**HAA (Height Above Airport)** The number of feet above the airport's published elevation when you reach MDA on a circling approach.

**HAT (Height Above Touchdown)** Your elevation above the touchdown zone of the runway when you reach the minimum altitude on a straight-in approach.

**high altitude** Refers to IFR operations at and above 18,000 feet.

**HIRL (High-Intensity Runway Lighting)** Spaced evenly down either side of the runway, these lights identify the edges of the paved surface and differ from ordinary runway lights in that the intensity can be increased or decreased to suit visibility conditions.

**HIWAS (Hazardous Inflight Weather Advisory Service)** Continuous recorded weather advisories, broadcast on selected VOR frequencies.

**holding** An orderly means of "retaining" an aircraft at some specified point by circling in a racetrack pattern.

**holding fix** The electronic point to which a holding aircraft returns on each circuit of the holding pattern.

**hypoxia** A physiological condition arising from the lack of sufficient oxygen to perform normal functions. Suffered in varying degrees by almost everyone when flying at altitudes above 10,000 feet.

**ident, squawk ident** Terms used by radar controllers when requesting pilots to activate the positive identification feature of a radar transponder. Causes the signal on a radar scope to take on distinctive characteristics for instant recognition by the controller.

**IFR** Stands for Instrument Flight Rules, but is used universally as a label for all instrument operations.

**ILS (Instrument Landing System)** A combination of electronic components which furnish information in all three dimensions (lateral, longitudinal, and vertical), designed to lead an aircraft to a missed approach point very close to the runway.

**ILS Category II** A more precise system that has even lower minimums than the normal, or Category I, Instrument Landing System. Requires special certification for ground equipment and airborne receivers, and more stringent pilot qualifications.

**IM (Inner Marker)** A radio transmitter identical to outer and middle markers, except for distance from the runway and signal pattern. IMs are situated between the MM and the runway, and usually transmit continuous dots (· · · · · · · · ·).

**IMC (Instrument Meteorological Conditions)** Weather conditions less than the minimums prescribed for VMC, or Visual Meteorological Conditions.

**in radar contact** A controller's confirmation that your flight has been positively identified on radar and will receive appropriate radar services until you hear "radar services terminated."

**jet routes** Airways in the high-altitude route structure (18,000 feet and above), such as J50 and J437. Should be referred to on the air as "Jay fifty," "Jay four three seven," and so forth.

**jet stream** A meandering river of high-speed air (sometimes moving at 200 knots or more), generally found at high altitudes. Like smart people up north, it usually moves south in the winter.

**jump** What you may want to do when all your radios and navigational equipment and instruments fail at night in a thunderstorm.

**kill the rabbit** A coded request for the tower controller to reduce the intensity of the RAILs (Runway Alignment Indicator Lights). It is usually heard on nights when visibility is very low and pilots are being blinded by the bright lights.

**kilometers** Used to quote visibility everywhere else in the world. It is designed to completely confuse all U.S. pilots accustomed to distances measured in feet and miles.

**knot** The expression of speed corresponding to 1 nautical mile per hour. Used universally by ATC and all foreign nations. Pilots are expected to accomplish—and refer to—all IFR operations in terms of knots.

**large** *See* aircraft classes.

**LBCM (Locator at the Back Course Marker)** A nondirectional radio beacon colocated with the fan marker on a back course localizer approach.

**LDA (Localizer-type Directional Aid)** Same as a localizer, but offset from the runway heading. Provides course guidance down to a point from which you can proceed to the airport by visual references.

**LDIN (Lead-in Lighting System)** A flashing (or distinctive) lighting system on the ground with light units in groups of at least three, positioned in a curved path to a runway threshold. Furnishes directional guidance only and should not be confused with ALS.

**LOC (Localizer)** The left-right information portion of an ILS; an electronic extension of the centerline of the runway.

**locator** Same as compass locator.

**LOM (Locator at the Outer Marker)** Same as compass locator.

**LORAN (LOng RAnge Navigation)** An electronic navigation system that uses hyperbolic lines of position from pairs of ground transmitters. LORAN is most useful on long, direct flight segments. It is more accurate than VOR, less accurate than GPS.

**low altitude** When referring to airway routes and charts for instrument operations, means "below 18,000 feet."

**low-altitude alert** A phrase used by controllers to get your attention when they observe a potentially hazardous proximity to terrain or obstructions. Usually followed by "check your altitude immediately."

**MAA (Maximum Authorized Altitude)** The highest usable IFR altitude in a specific airway segment. Always clearly indicated on IFR enroute charts.

**maintain** What you are expected to do when assigned an altitude by ATC.

**MAP (Missed Approach Point)** Expressed in either time or distance from the final approach fix or as an altitude on the glide slope. It is the point at which a missed approach must be executed if the runway environment is not in sight.

**map** Usually referred to by pilots as a "chart."

**marker beacon** Same as fan marker.

**marker beacon lights** Two, or sometimes three, panel-mounted lights that illuminate appropriately to indicate passage over radio range markers on an instrument approach.

**MAYDAY** The universal code for an emergency. To be used only in dire circumstances to get the attention of anyone who might be listening. Normally, an emergency condition can be brought to the attention of ATC (or others) by using plain language to explain the problem.

**MCA (Minimum Crossing Altitude)** Indicated by an "X" flag on enroute charts when a certain altitude must be attained prior to going beyond an airway intersection.

**MDA (Minimum Descent Altitude)** The lowest altitude (expressed in feet above sea level) to which you may descend on a nonprecision approach (one without a glide slope) if the runway environment is not in sight.

**MEA (Minimum Enroute Altitude)** The lowest altitude at which you can receive a satisfactory navigational signal for the appropriate segment of a federal airway.

**minimum fuel** When declared to a controller, indicates that you can make it to the airport with normal handling, but that any delay may result in your declaring a full-blown low-fuel emergency.

**missed approach** A procedure specified for a "go-around" if the runway environment is not in sight at the missed approach point.

**MM (Middle Marker)** A highly directional radio beacon located about one-half mile from the end of the runway on an ILS approach. A distance indicator, it transmits a signal of high-pitched alternate dots and dashes ($\cdot — \cdot — \cdot — \cdot —$). If marker beacon lights are installed in the airplane, the amber light will flash in a similar pattern. The middle marker does not *always* indicate the missed approach point.

**MOA (Military Operations Area)** A volume of airspace that is indicated on all aeronautical charts and is used at various times and altitudes by the military services. IFR flights are automatically cleared through or rerouted; VFR pilots must exercise great caution when flying through these areas.

**MOCA (Minimum Obstruction Clearance Altitude)** Guarantees terrain and obstacle clearance on the appropriate segment of a federal airway. If you should have to fly at MOCA, remember that a usable VOR signal is guaranteed only within 22 miles of the station.

**MRA (Minimum Reception Altitude)** The lowest altitude at which you will be guaranteed reception of an off-airway VOR for the purpose of identifying an intersection. Has nothing to do with communication reception.

**MSA (Minimum Safe Altitude)** The lowest altitude within 25 nautical miles of an approach fix that guarantees 1000 feet obstacle clearance. Found on the approach and landing chart, it may be one altitude for all directions, or may be referenced to several sectors around the fix.

**MVA (Minimum Vectoring Altitude)** The lowest altitude (in a specific area) a controller can assign to guarantee terrain and obstacle clearance. MVA does not appear on pilot charts.

**nautical mile** One minute of latitude (measured vertically on charts). All distances on IFR charts are indicated in nautical miles; all onboard navigation systems display distance in nautical miles.

**NDB (NonDirectional Beacon)** A low-frequency radio transmitter that emits a signal in all directions from the antenna. Its signal is not aimed in any one direction, hence the term "nondirectional."

**negative** An emphatic, easily understood "no."

**no-gyro approach** A technique used by controllers when a pilot is operating in a "partial panel" situation—needle, ball, and airspeed.

**no joy** A term borrowed from the military; means "I do not see the traffic you told me about," and saves a lot of communications time.

**nonprecision approach** A published procedure (or surveillance radar approach) which does not provide an electronic glide slope. All approaches except ILS (and military precision radar) are nonprecision.

**NOTAM (Notices To AirMen)** Information pertinent to operations in the National Airspace System.

**obscuration** Weather condition in which the sky is hidden by ground-based phenomena such as rain, fog, snow, etc. The height given in such an observation is vertical visibility only; forward vision will usually be severely restricted.

**OM (Outer Marker)** Usually the final approach fix on an ILS or localizer approach, this 75 megahertz, nontunable radio beacon is typically located 4 to 7 miles from the runway. Highly directional, its signal is transmitted in a narrow beam straight up and is received only when directly overhead. The audible signal is a continuous series of low-pitched dashes (— — — — —), and if marker beacon lights are installed in the airplane, the blue light will flash in conjunction with the sound.

**omni** Short for Very high frequency Omnidirectional Radio range, also known as VOR.

**outer locator** Same as LOM.

**over** Used at the end of a radio transmission to indicate that a reply is expected. Quite unnecessary—if what you have said requires a reply, you'll get it anyway.

**over and out** Used mostly by the Walter Mitty types who also wear helmets and goggles in their Learjets.

**PAR (Precision Approach Radar)** Provides a precision approach based on vocal instructions given to the pilot, as the controller observes azimuth, distance, and elevation on radar. Known as GCA (Ground Controlled Approach) at military airfields. PAR is now almost nonexistent in the civilian system.

**parallel ILS approach** In existence at only a handful of airports, and operable only within very restrictive conditions of runway separation and radar monitoring, these approaches provide guidance to side-by-side runways simultaneously.

**partial panel** Term used to describe the situation that exists when the attitude indicator and heading indicator are inoperative, or covered up with some fiendish device your flight instructor pulls from his shirt pocket. Sometimes referred to as "needle, ball, and airspeed," especially among gray-haired pilots.

**PIREP (PIlot REPort)** Weather information provided by pilots in flight. PIREPs are transmitted throughout the weather-reporting system and are available to other pilots for the asking.

**precipitous terrain** Where steep and abrupt slopes exist under the final approach course, the approach chart will display a note to that effect. In addition to causing violent up and down drafts, the sudden changes in pressure can result in erroneous readings of altitude and airspeed. Approach planners take this into account, and boost the minimum altitudes accordingly.

**precision approach** An approach procedure which incorporates an electronic glide slope. In today's IFR world, there is only one in regular use, the Instrument Landing System (ILS).

**preferred IFR route** The way ATC would like you to file between specified airports. Especially in congested areas, the preferred route is probably the one you'll get, no matter what you file.

**prevailing visibility** The horizontal distance at which known objects can be seen through at least half the observer's horizon. It does not necessarily mean a *continuous* half of the horizon, and so can vary considerably from the actual visibility at the approach end of the runway. Other types of visibility measurement (e.g., RVR) are considered more appropriate for instrument approaches; of course, the pilot's observation is the ultimate.

**procedure turn** A means of reversing course to line up inbound on the approach course. Generally used in nonradar situations or when you arrive over the approach fix headed away from the airport.

**procedure turn altitude** Found on the profile view of the approach chart, this is the minimum altitude while reversing course. You are considered in the procedure turn and must observe this altitude limit until once again on the approach course inbound.

**Queen, Hangar** What you call your airplane when it's in the shop more than it's in the air.

**radar (RAdio Detection And Ranging)** Pulses of electronic energy are emitted from a ground transmitter, and objects that reflect the energy show up on the radar scope as "blips" of bright light. Azimuth, distance, and in some cases altitude can be measured. Airborne transponders cause a coded blip to appear on the scope.

**radar beacon** The official name for a transponder.

**radar contact** Phrase used by controllers to indicate that a signal representing your aircraft has been identified on the radar scope.

**radar-monitored approach** One in which a radar operator follows the progress of an aircraft in order to provide corrections in course and, sometimes, elevation. Used mostly at very congested terminals. You can request a radar-monitored approach in an emergency or when you suspect navigational equipment malfunction.

**radar service terminated** A phrase used by air traffic controllers to inform pilots that whatever radar services (including traffic advisories) they have been enjoying are hereby withdrawn.

**radial** A magnetic course radiating, or proceeding outward from, a VOR. Controllers *always* use the term "radial" when referencing navigation instructions to a VOR.

**radio altimeter** An electronic device that measures the time for a radio signal to go from the aircraft antenna to the ground and return; it then displays the time interval as a precise distance. Used generally for precision approaches and is required for Category II and Ill ILS operations. (Frequently referred to in error as a "radar" altimeter.)

**radio compass** Same as ADF.

**RAIL (Runway Alignment Indicator Lights)** The "ball of fire" you see running toward the runway in a fully equipped approach lighting system. Sometimes called the "rabbit."

**read back** What a controller sometimes says at the end of an IFR clearance. It's good practice to read back at least altitude assignments and vector headings, and in the sequence issued.

**REIL (Runway End Identifier Lights)** A pair of strobe lights identical to those in the RAIL system, but located on either side of the runway threshold. REILs flash simultaneously, about once a second.

**relative bearing** On an ADF indicator, the number of degrees the pointer is displaced *clockwise* from the nose (top index) of the aircraft. Add relative bearing to magnetic heading, and the resulting number is *always* magnetic course to the station.

**report landing assured or missed approach** When cleared for an approach to an airport with no ATC facility, you will be separated from other IFR traffic. Should you be required to execute a missed approach, you go right back into the system for another try or perhaps clearance to your alternate. The controller would therefore like to hear from you when "landing is assured" (runway in sight) or when you start a missed approach.

**report procedure turn** A request for a report as soon as you turn away from the outbound course; in other words, at the very beginning of the procedure turn itself.

**report procedure turn inbound** When so instructed on an instrument approach, the controller wants to hear from you as soon as you are back on course, headed toward the airport.

**resume normal navigation** Usually follows "radar service terminated" or vectors for traffic, and is your cue to intercept, stay on, or get back to the appropriate radial, airway, or course. You're then expected to get where you're going all by yourself.

**RMI (Radio Magnetic Indicator)** A slaved gyro that displays aircraft heading at the top of a rotating card, with pointers tied electronically to the VOR and/or ADF receivers. When properly tuned, the pointers indicate magnetic course to the appropriate station. The RMI is also convenient to use as a heading indicator.

**RNAV area navigation** Refers to the use of any onboard electronic system that processes information from VOR-DME stations, LORAN transmitters, GPS satellites, or other means of navigating within a specific area (hence the name). An RNAV system automatically and continuously computes the course and distance to a desired position, called a "waypoint."

**roger** Worldwide aviation term that indicates complete understanding of a radio transmission. Can also be used (with appropriate volume) when answering a bartender to let everyone in earshot know that you are a pilot.

**roger wilco** Means "yes, I understand and *will* comply," and is completely out of place in today's aeronautical communications. A simple "roger" or your aircraft call sign will do the job.

**runway environment** The runway threshold, approved lighting aids, or other markings identifiable with the runway. Having the runway environment in sight is one of the requirements for descending below DH or MDA on an instrument approach.

**runway heading** What you're expected to fly when cleared to "maintain runway heading" as part of a takeoff clearance. Don't correct for wind; fly the heading as published.

**RVR (Runway Visual Range)** A photoelectric device called a "transmissometer" measures the visibility immediately adjacent to the runway, then computes what a pilot would see, and converts this observation to hundreds of feet. Displayed in the tower, it will be volunteered by the controller or issued at your request in low-visibility conditions and takes precedence over all other forms of visibility measurement, except your own two eyes.

**say again** Please repeat all of what you have just said.

**say again all after...** Please repeat everything that followed a given word or phrase—implies complete understanding of whatever came before that point in the transmission.

**say again all after ATC clears...** Generally used by honest pilots when they fail to copy an IFR clearance.

**say again, you broke up** This is a good way to "keep your cool" when you don't have *any* idea what the controller said—no one can prove that the transmission *didn't* break up!

**SDF (Simplified Directional Facility)** An instrument approach procedure that fits on the accuracy scale between a VOR and a localizer approach. Course width may vary from 6 to 12 degrees among installations, with the final approach course offset somewhat from the runway heading.

**SID (Standard Instrument Departure)** A published clearance procedure designed to cut down communication time. Each SID is named and numbered, and directs you to an enroute fix from which you can proceed via the routing in your clearance.

**sidestep** When approaching an airport with parallel runways very close together, the tower will occasionally clear an aircraft to "sidestep" and land on the other runway. This must always be a visual maneuver.

**SIGMET (SIGnificant METeorological advisory)** A weather advisory that makes pilots aware of conditions such as turbulence, severe icing, and the like. A SIGMET contains information that may affect the safety of all aircraft. *Convective* SIGMETs are advisories of severe thunderstorm conditions.

**small** *See* aircraft classes.

**special VFR** An ATC clearance (sometimes called "poor man's IFR") to proceed into or out of Class C or D airspace under certain conditions.

**squawk altitude** Phrase used by controllers when they want you to adjust the transponder to transmit an altitude signal.

**squawk standby** When so directed, turn the transponder function switch to "standby," which will cause the transponder signal to disappear completely from the radar scope. Used for positive identification when the IDENT feature is inoperative or weak. Usually followed by "squawk normal" when the controller makes a positive identification.

**standard rate turn** A turn which results in the aircraft changing heading at the rate of 3 degrees per second. To approximate the angle of bank required, divide true airspeed by 10 and add 5.

**stand by** The universal reply to a question when the answer is not readily available.

**STAR (Standard Terminal Arrival Route)** A published route into a terminal area, with the basic aim of reducing communications between pilots and controllers. Each STAR has its own name and number; for example, "cleared via the Shirt Four Arrival."

**star** A bright spot in the sky, sometimes mistaken for the lights of another airplane.

**stepdown fix** An intermediate point between the Final Approach Fix and the Missed Approach Point. A lower altitude is permissible after you pass the stepdown fix.

**stop altitude squawk** Phrase used by controllers when, for reasons of inaccuracy or internal problems, they want you to remove the altitude information from your transponder signal. To comply, return the function switch to "Normal."

**straight-in approach** A procedure which leads directly to the landing end of the runway, when that runway is aligned no more than 30 degrees from the final approach course and a normal descent from the minimum IFR altitude (MDA) can be accomplished.

**surveillance approach** A nonprecision procedure in which approach control radar is used to position an aircraft on the final approach course and guide the pilot to the missed approach point. Usually a "semiemergency" situation when on-board navigational equipment or flight instruments have failed or malfunctioned.

**TACAN (TACtical Air Navigation)** A military navigation system combined with a VOR to make a VORTAC, which supplies both azimuth and distance to DME-equipped civilian aircraft.

**tallyho** Say this when you spot the traffic that Center points out to you as in, "1234 Alpha, traffic at 2 o'clock, 4 miles, eastbound." Tallyho is a bit melodramatic, but it sure cuts down on communication time!

**TAS** True AirSpeed.

**TDZ (TouchDown Zone)** Extends 3000 feet down the runway from the threshold or the first half of the runway, whichever comes first. The highest elevation in the TDZ provides the point from which DH and MDA are measured for straight-in approach minimums.

**track** (1) The path which an aircraft makes across the ground. When cleared direct from one point to another, your track is expected to be a straight line. (2) Parallel lengths of ferrous material secured to wooden crossties and used for aerial navigation. Also used now and then by long lines of railroad cars.

**transmissometer** The photoelectric visibility-sensing device that provides the basic input for RVR measurements. It is installed beside the landing runway at the approach end and, at some airports, at midfield and roll-out locations as well.

**transponder** An onboard electronic device which, when interrogated by a ground radar signal, responds with a coded return, providing a distinctive display on controllers' screens for positive radar identification.

**transponder code** The specified sequence of numbers which, when selected on the transponder, provides a coded display on ATC radar scopes.

**UCT (Universal Coordinated Time)** Based on the time at Greenwich, England (site of the Queen's observatory), UCT is the time standard used throughout the world. Previously known as Greenwich Mean Time, also code-named "Zulu" time.

**unable** The communications code word used to indicate your inability to contact a facility; for example, "Tower, 1234 Alpha *unable* Approach Control on 123.8." The tower controller will then give you further instructions.

**vector** A heading assigned by a radar controller that takes precedence over all other forms of navigation. You are expected to remain on vectored headings until advised "resume normal navigation."

**verify** When used by either a controller or a pilot, means "did you really say what I think you said?"

**VFR (Visual Flight Rules)** Conditions to which pilots are limited until they become instrument rated. A nonspecific term compared to IFR, which means "I Follow Railroads."

**visibility** What you need a certain amount of in order to legally land at the end of an instrument approach. It may be measured and reported by a ground observer, a transmissometer (RVR), or the pilot. *You* become the official visibility observer on an approach—if the runway environment is in sight at a certain point, continuing to a landing is completely legal.

**visual approach** A shortcut to a published instrument approach procedure, in which you are cleared to proceed direct to the airport or to follow another aircraft headed for the same field.

**VOR** Stands for Very high frequency Omnidirectional Radio range.

**VOT** Stands for VOR test facility, with which you can check the accuracy of onboard VOR indications either in the air or on the ground.

**waypoint** A geographical position determined by any onboard RNAV system, used for enroute navigation as well as approved IFR approach procedures.

**What am I doing here?** The question you will ask yourself when flying in the middle of a thunderstorm.

**Wilco** I have received your message, understand it, and WILl COmply with it.

**X-ray vision** What ATC expects you to have when the controller says, "The airport is at 1 o'clock, 2 miles; do you have it in sight?"

**Zulu time** A military-spawned code word which is easier to say than "universal coordinated time." Means the same thing and is frequently shortened to "Z."

# 3

# Attitude Instrument Flying

The pterodactyl, a lizard turned bird that soared on huge leather-skinned wings above the swamps and marshes of prehistoric times, may well have been the first creature to experience IFR flight. Because the earth was frequently covered with clouds of noxious gases from the volcanoes that dotted the landscape, it's a good bet that the pterodactyls on occasion ignored their VFR-only limitations and flew into IFR conditions. Of course, no one was there to prove it, but quite likely they had an instinctive reaction to counter the sudden loss of visual clues; they were probably able to set their wings for an optimum glide speed, maintain direction with some primitive vestibular gyroscope, and keep going until they broke out of the clouds. It's just as likely that some of them didn't make it; this theory is supported by the discoveries now and then of pterodactyl remains on or just below the summits of mountain peaks.

The blind-flying capabilities of our feathered friends extend to many of the birds we know today—ducks, pigeons, geese, and others have been known to successfully navigate through the clouds to safety. Humans have not fared so well, requiring some sort of help from artificial references to maintain their spatial orientation. But anyone who has flown Cubs, Airknockers, and the like knows that if things really get bad, they could always close the throttle, roll the trim all the way back, keep the turn needle centered with the rudder, and at least come out of the situation right side up.

When aviation pioneers turned their efforts to making money with airplanes, they soon realized that they would have to figure some way to overcome what appeared to be the insurmountable problem of flying all the way through clouds — halfway just wouldn't do. A frightening number of our departed predecessors tried it on guts and confidence, but it all boiled down to the necessity of substituting some kind of instrumentation to replace the natural horizon when it disappeared. Early attempts included "playing it by ear," listening to the sound of the wind in the ample wires and stays that held those old airplanes together, watching the flutterings of pieces of cloth tied to the struts, and even Mason jars half-filled with oil that were supposed to indicate whether the airplane was banking or in level flight.

None of these early methods worked very well, but with the advent of gyroscopic devices, a whole new world of airplane utility was introduced. When Jimmy Doolittle proved that a pilot could take off, navigate, and land an airplane using no outside visual references, he introduced a method of instrument flight that we use essentially unchanged to this very day.

Whether you took flying lessons yesterday or 30 years ago, one of the first things you learned was that when the curve of the engine cowling, or the top of the instrument panel, or some other reference showed a certain relation to the horizon, and when you had the engine wide open, the airplane would climb at a predictable airspeed and vertical speed. You learned that when, at a given airspeed, the airplane was banked until the center post of the windshield formed a particular angle with the horizon, the airplane turned at a certain rate. The substitution of a miniature airplane fastened to the instrument case and moving about a gyro-stabilized bar in the same direction and to the same degree as the actual movement of the real airplane merely transfers to an artificial horizon what you used to see outside. If you set up your old familiar climb attitude with outside references and at the same time notice what you see on the attitude indicator, you can forget the outside clues and rest assured that anytime you put the miniature airplane in the same place, with the same power setting, you're going to get the same old familiar climb performance. That, in a very small nutshell, is attitude instrument flying. It's basic, reliable, and really very simple to accomplish.

But the attitude indicator can't do the job alone. Although it's the heart of the system, you must refer to the other instruments to determine what is happening when you select a pitch attitude that's 3 degrees above the horizon, or a 20-degree bank, or a combination of the two. In other words, you must reason to yourself, "I will place this little flying machine in the attitude that I *think* will produce a standard-rate climbing turn to the right, and then check the other gauges to see if I was right. If the needles and pointers aren't moving the way they should, I will change the attitude a bit to get the results I want." It's easy to see that, if all the variables are held constant, the same attitude will give the same condition of flight every time. Now, in a manner of speaking, you *can* set your wings just like the birds and have confidence that certain things will take place.

# ✝ The Fundamentals of Attitude Control

Pitch, bank, power, and trim...the fundamentals of attitude instrument flight. When you get right down to it, there's not much you can do with an airplane except change the pitch, bank, or power, and how much you change what in which direction in concert with the others or all by itself determines what will happen. You may feel that this listing of the basics added trim unnecessarily and left out rudder control; not so, because trim is of great importance in smoothing out your cloud-bound wanderings, and you are expected to maintain the ball in its centered position at all times in instrument flight. In general, you should always adjust pitch, bank, and power, then trim to hold the airplane in the attitude you want. The human machine is incapable of holding a constant, precise pressure for an extended period of time, but it can *recognize* that pressure and adjust trim controls until it disappears.

Now, you must become a believer. Raise your right hand, and repeat for all to hear: "I [state your name] do hereby solemnly affirm my belief in the fact that in all steady-state instrument situations (that is, whenever a constant airspeed, constant altitude, or constant rate of change of altitude is desired) altitude shall be controlled by *power* and airspeed shall be controlled by *pitch*." It's true! Believe!

To prove it, go back in memory to that training exercise known as flying at "minimum controllable airspeed." Once you were established at an airspeed very close to stalling and the altitude began to droop a bit, you didn't pull up the nose to correct; you added *power*. If the airspeed was somewhat higher than you wanted, you didn't back off on the throttle; you applied *back pressure*. And it works just as well when you're under the hood or in the clouds—perhaps even more so than when you can see outside because you have an exacting reference right there in front of you. It's called an "instrument panel."

When that gaggle of needles, pointers, and gauges is complete and everything's working properly, one instrument is primary—the attitude indicator (AI). It seems reasonable, in a context of attitude instrument flying, that the attitude *indicator* should be the most frequently consulted because it tells you more at a glance than anything else on the panel. Changes in attitude on this instrument will be immediately reflected in the readings elsewhere; you are able to make changes or keep them from taking place by referring to the AI. Anytime it's not in the level flight attitude for the particular power setting in use, something is happening, or is about to happen. (Probably 99.9 percent of your instrument time will be spent scanning the indications of a so-called full panel, with all the instruments in working order and doing their jobs. However, mechanical devices being what they are, it's possible someday to find yourself operating in a "partial panel" situation, and you should be able to recognize it and know how to handle the airplane without a full complement of instruments. See Chap. 4 for details.)

Your first step in becoming a good attitude instrument pilot is to find out what the gyro horizon looks like when you are flying straight and level. Nearly all indicators in use today are adjustable, so you can move the miniature airplane up or down so that it is dead center on the artificial horizon in a straight and level, constant airspeed condition. Once you have found this attitude and adjusted the little airplane to show it, *leave it there.* This gives you a sound base from which to proceed. Now you can change the attitude as required—for a constant airspeed climb at full power, you may have to increase the pitch attitude two bar widths (raise the miniature airplane twice the thickness of its wings above the horizon bar), or perhaps you will recognize that a one-half bar width increase is required to hold altitude in a standard rate turn. Every time you make that little airplane move, the movement should be referenced to the level flight attitude as shown on the attitude indicator. (Slight adjustments for varying loads and atmospheric conditions may be required.)

Sources of qualitative information will change in various maneuvers. For example, when rolling into a standard rate turn, bank the miniature airplane until the turn needle moves to the proper indication and then maintain the rate of turn as shown on the needle by making minor changes in the bank attitude. The turn needle has become the primary source of information about rate of turn, but the attitude indicator remains stalwart in its indication of the bank required to sustain this condition. As you approach a predetermined heading, the heading indicator will tell you when to begin the roll-out, and it's back to the attitude indicator to return the airplane to level flight. When you're trying to maintain a precise rate of descent at a given airspeed (as in the final segment of a nonprecision approach), the airspeed indicator tells you whether or not the attitude you have selected is the proper one, and the attitude indicator is used to make any changes that may be required. Since you believe that power controls altitude (and altitude change), when it is necessary to increase your rate of descent, you will reduce power with reference to the engine instruments, but you need the attitude indicator to maintain the pitch relationship that will keep the airspeed constant. And of course, whenever straight flight is your target, keeping the wings level with the attitude indicator (and the ball centered as always with rudder pressure) can do nothing but make the airplane fly straight ahead.

# ✝ Applying the Attitude Technique

Some instrument pilots are sharper, smoother, and more precise than others—that's a fact of life and is due partly to the physical and mental capabilities of each individual. But running through the performance of even the slowest, most plodding IFR aviator is an undeniable 1-2-3 sequence of events. Some are able to accomplish it faster than others, but you can't get away from (1) observing the instrument indications, (2) figuring out what they have to say to you, (3) doing something about it. This sequence is commonly known as cross-check, interpretation, and control.

## Instrument Cross-Check

Tirades from flight instructors, grueling sessions in ground trainers, and sophisticated eye-movement studies have failed to come up with a formula for cross-checking flight instruments. In addition to the fact that the amount of time spent on any one instrument must change subtly with every change in flight condition, cross-checking seems to be a very personal matter. The number of times per minute your eyes scan the entire panel, or how long they remain on the turn needle or the heading indicator, isn't really important; the essence of cross-checking lies in the *amount* of information gleaned from each sweep of the gauges. You must develop your cross-check to provide enough inputs for meaningful decisions about what to do next, if anything. It stands to reason that the more complicated the maneuver, the faster your eyes must move, especially when you are introducing a change of attitude.

There is only one way to become a proficient cross-checker; force yourself to eyeball each of the instruments, with particular emphasis on the attitude indicator (you'll be amazed at the minute changes you can detect), and practice, practice, practice. Here's where the proficiency exercises described in Chap. 21 can be a big help; without the distractions of Approach Control and with no navigational load to divert your attention, go through these maneuvers if for no other reason than to speed up your cross-check. It's a skill, and as such it must be practiced regularly and methodically if it's to stay sharp.

Later on, you must begin to include navigational inputs in your cross-check. It's easy to develop instrument hypnosis characterized by ignorance of nav-aid information. Pilots have been known to fly all the way to the missed approach point with great precision, but at procedure turn altitude—dead center, on course, and so concerned with maintaining an exact altitude they forgot to descend over the Final Approach Fix.

# ✝ Instrument Flying for Animal Lovers

Having detailed the concept of attitude control, there is another method that you may prefer. For reasons that will become apparent, it is recommended for those pilots whose airplanes have large, easily cleaned cabins. Known as the "Cat and Duck Method," or C&D Method, of instrument flight, it has received a lot of publicity and is considered to have a great deal of merit by those who have not tried it. No reports have been received from those who did try it, and none are expected. You are invited to assess its merits objectively.

Basic rules for the C&D Method of instrument flight are fairly well known and are extremely simple. Here's how it's done:

1. Place a live cat on the cockpit floor; because a cat always remains upright, it can be used in lieu of a needle and ball. Merely watch to see which way the cat leans to determine if a wing is low and, if so, which one.

2. The duck is used for the instrument approach and landing. Because of the fact that any sensible duck will refuse to fly under instrument conditions, it is only necessary to hurl your duck out of the plane and follow it to the ground.

There are some limitations to the Cat and Duck Method, but by rigidly adhering to the following checklist, a degree of success will be achieved which will surely startle you, your passengers, and even an occasional air traffic controller.

1. Get a wide-awake cat. Most cats do not want to stand up at all. It may be necessary to carry a large dog in the cockpit to keep the cat at attention.

2. Make sure your cat is clean. Dirty cats will spend all their time washing. Trying to follow a washing cat usually results in a tight snap roll followed by an inverted spin.

3. Use old cats only. Young cats have nine lives, but old, used-up cats with only one life left have just as much to lose as you do and will be more dependable.

4. Beware of cowardly ducks. If the duck discovers that you are using the cat to stay upright, it will refuse to leave without the cat. Ducks are no better in IFR conditions than you are.

5. Be sure that the duck has good eyesight. Nearsighted ducks sometimes fail to realize that they are on the gauges and go flogging off into the nearest hill. *Very* nearsighted ducks will not realize they have been thrown out and will descend to the ground in a sitting position. This maneuver is difficult to follow in an airplane.

6. Use land-loving ducks. It is very discouraging to break out and find yourself on final for a rice paddy, particularly if there are duck hunters around. Duck hunters suffer from temporary insanity while sitting in their blinds in freezing weather and will shoot at anything that flies.

7. Choose your duck carefully. It is easy to confuse ducks with geese because many water birds look alike. While they are very competent instrument flyers, geese seldom want to go in the same direction as you. If your duck heads off for Canada or Mexico, you may be sure that you have been given the goose.[1]

---

1. I wish I could take credit for this amusing treatise on instrument flying, which I first saw in a Naval Aviation safety publication years ago. But alas, the original author was none other than, and remains to this day, Anonymous.

# 4

# Partial Panel IFR

Most of us tend to be complacent about the performance of things mechanical, especially when those things have worked properly for a long period of time. That is surely the case with most instrument pilots and their respect for the artificial horizon and directional gyro—"Sure, I've heard of an instrument failure now and then, but mine seems to be okay; no cause for alarm."

But most instrument failure accidents suggest strongly that the average instrument pilot needs to be very aware of the potential for a gyro failure, and the *probability* of his losing control of the airplane in instrument conditions unless he's properly trained, prepared, and current. Whether the problem is caused by a malfunction of the instruments themselves or by a failure of the vacuum/air pressure source (the most likely event), the result is the same—a life-threatening situation that requires seldom-practiced skills and techniques for survival. The pilot community has dubbed this unhappy circumstance "partial panel" flying.

## ✈ The Heart(s) of the Problem

Up to this point, we've emphasized the principle of "attitude flying," wherein a certain attitude coupled with a certain power setting will always produce a certain type of performance. That technique is based on an operating attitude indicator, which provides a clear and precise picture of airplane attitude in two axes—roll and pitch.

But when the attitude indicator fails, you're playing with a different set of rules. It's not really attitude flying anymore because all you have to work with are the *reactions* of pitch, power, and bank; there is no longer anything on the panel that shows attitude. The remaining flight instruments (turn indicator, altimeter, airspeed indicator, vertical speed and magnetic compass in a typical lightplane installation) show you what's happening, but they provide no *direct* indication of how high the nose has been raised or how much the wings have been banked. Since you are now working with results only, a different technique is a *must* if you are to fly safely out of this bad situation.

The first "heart" of the partial panel problem is the very recognition of something amiss. Of course, you can keep pushing the problem farther into the future by replacing or repairing any flight instrument that exhibits symptoms of impending failure. If an attitude indicator or directional gyro (AI and DG from now on) takes forever to stabilize after engine start or makes weird, grinding noises as it winds down at the end of a flight, call in the instrument doctor and find out what's wrong. Better to spend the bucks for a new or rebuilt instrument than to put yourself through a partial panel wringer for real on some future IFR flight.

Unfortunate but true, gyro instrument failures are likely to be rather benign in character. The usual scenario goes like this: The power source fails after the gyros are running at full speed, whereupon the instruments *gradually* lose their ability to tell you the truth. The AI is the prime culprit here; it will probably roll into a bank at such a slow rate that you'll follow right along with it, and by the time you realize you've been had, you've been had! This is the reason why every thorough IFR flight check should include an evaluation of the pilot's ability to recover safely from at least two abnormal attitudes— a climbing turn and a diving turn (see Exercise #12, Chap. 21).

You can maintain a constant check on the health of your air-powered flight instruments by including the vacuum/pressure gauge in your normal scan of engine instruments. You might also install a big red warning light to get your attention when the value falls below an acceptable level—easier said than done, of course, but there's no good reason *not* to maintain a close watch on this Achilles' heel.

The other preventive measure is to constantly check all of the flight instrument indications against each other. If the AI shows a bank to the right, is the DG increasing? If the DG numbers are moving to the left, is the turn needle deflected the same way? Do the pitch attitude and the altimeter reading make sense? Whenever instrument indications disagree, it's time to check the power source and then prepare yourself to shift gears; it may well be partial panel time.

The second "heart" of this problem is what to do after you've determined that all is not well in the flight instrument department. From now on, we're concerned only with the techniques and procedures applicable to an attitude instrument failure in actual IFR conditions. You don't want to be there, but you *are,* and there's only one way out: Fly the airplane someplace where you can see the ground, find an airport, and land safely. Keep that firmly and foremost in mind because your survival depends on your ability to take command and *fly the airplane.*

Experience has shown that partial panel operations in the clouds represent a true emergency for most pilots. There is assistance available from air traffic controllers (they will direct you to a suitable airport, help you find VFR weather conditions, or perhaps offer nothing more than an occasional "you're doing fine!" to keep your spirits up), but if they don't know, they can't help. With that in mind, declare an emergency as soon as you recognize a partial panel situation. There's no need to pick up the mike and scream "MAYDAY!" but be sure the controllers understand clearly that you have a real problem and need all the help you can get. The mere mention of the word "emergency," in calm, quiet tones, will get the attention you deserve. Don't be too proud to solicit additional inputs; the controller might have the secret to your survival.

## ✛ Problem Identified, Now What?

If you've gotten yourself into an unusual attitude by the time you realize a problem exists, the recovery procedures in Exercise #12 will save the day; but without the AI and DG, you must change your technique a bit. Add or reduce power according to your assessment of airspeed and then simultaneously center the turn indicator *with aileron pressure* and center the ball *with rudder pressure.* This will virtually guarantee a coordinated return to straight flight.

Now you must take care of the pitch attitude. If you were doing a good job of flying before the instruments went haywire, the airplane was well trimmed and should therefore be trying to regain the trim airspeed. You probably won't be able to sit there and wait, so an adjustment of pitch attitude is in order. Without the AI to show pitch changes, you'll need to make an actual movement of the yoke and then evaluate the results.

Bring the altimeter into play at this point; if it is decreasing, physically move the yoke aft a small amount—perhaps a quarter-inch or so—and note any change in the rate at which the altimeter is moving (the vertical-speed indicator may be of some help here, but the inherent lag in its indication will likely cause you more trouble than it's worth). If there's no change, move the yoke back another quarter-inch; keep this up *until the altimeter stops moving*—you are now in level flight (assuming that you have kept the turn needle and the ball centered all the while).

Slight changes in pressure on the yoke will now permit a return to the stabilized airspeed for which the airplane was trimmed, and once achieved, *do not change the trim.* You have arrived in a known, flyable condition, one that will get you where you need to go. Don't upset the apple cart by changing the trim.

At this point, recognize that, without direct readings of airplane attitude, the fewer the control inputs the better. So take your hands off the wheel, relegate directional control to the rudder, and altitude changes to the throttle. You'll find that gentle, steady pressure on the appropriate rudder pedal will start and maintain a shallow turn or will hold the heading for straight flight. The turn indicator becomes primary for directional control now. Keep it centered to maintain a heading; move it left or right for turns.

Altitude control is vested in the throttle. Because level flight results from just enough thrust being applied to keep the airplane from climbing or descending at the trimmed airspeed, it stands to reason that adding power will result in a climb at the same airspeed, whereas reducing power will set up a descent. You'll need to experiment with your airplane to find out how much power change creates a reasonable climb or descent.

In summary, when on partial panel with the airplane trimmed and stabilized, all heading changes will be made with rudder pressure, and all altitude changes will be made with power. It's the best way to keep from flying yourself right back into that unusual attitude you worked so hard to defeat. You may even conclude that hands-off instrument flying works so well that you'll adopt it as your normal technique.

Spatial disorientation is but a heartbeat away from any pilot in a partial panel situation, especially when a really hairy unusual attitude is encountered. There will be noises and sensations you've not experienced before, and your inner ear will be undergoing gyrations that are all but guaranteed to turn you inside out. If ever you have rallied your resources to overcome inner conflicts, this is that time. Center the turn indicator, get the altitude and airspeed under control, and *cover up the AI and DG* to focus your attention and prevent any more distracting visual inputs. Isn't that the way you were trained to cope with the loss of those instruments?

Your instructor may have used a soap holder or a business card to take the AI and DG out of your scan, but there was still a requirement to apply common sense and good judgment to resolve the partial panel situation and get the airplane safely on the ground. In a real-world example several years ago, a CFI and student on a cross-country flight experienced a *partial* partial panel event when the DG rolled over and died. They were in the clouds at 4000 feet when it happened, and ATC granted their request to return home and land.

The controller provided radar vectors for a VOR-DME approach, but along the way, the CFI requested a descent to 1000 feet to get into VMC. Despite being warned that 1000 feet was well below the minimum IFR enroute altitude, the instructor persisted, and radar contact was lost shortly thereafter.

The airplane was found at the end of a long wreckage path, which began in the tops of high trees. It appeared the C-152 hit the trees wings-level, in a gradual descent at cruise airspeed. The instructor's soap holder was found stuck to the face of the failed DG.

The loss of primary flight instruments is indeed a bona-fide emergency, but it doesn't mean that basic judgment should fail at the same time. If the cloud base is below the minimum IFR altitude all the way home (or the nearest suitable airport), resign yourself to staying at a safe altitude until you can break out, or complete an instrument approach. Scud-running on partial panel is every bit as dangerous as it is with a full complement of instruments.

## ✛ Which Way Salvation?

Unless your partial panel problems cropped up on a perfectly aligned final approach to an airport, you'll have to change heading occasionally to get on the ground. The magnetic compass is suddenly worth its weight in gold, but there are serious, dangerous problems involved with its use in lieu of a DG. You were taught way back in private-pilot ground school that there is only one condition in which the magnetic compass is reliable—straight and level, unaccelerated (no airspeed change in progress) flight. There are methods of accounting for compass errors, but they are not suited for use in this situation.

An easier and far more reliable method of changing heading on partial panel is the timed turn. Get good at this, and you can forget about compass errors while you're turning; the so-called "whiskey compass" becomes nothing more than a device for confirming that what you have done is correct. This technique is based on the standard-rate turn, or a heading change of 3 degrees per second; all turn indicators are marked to indicate when that rate of turn is achieved. If your goal is to change heading to the right 30 degrees, note the reading on the mag compass (to provide a starting and ending point), then apply smooth, gentle pressure on the right rudder pedal until the turn indicator shows a standard-rate turn, and at the end of 10 seconds, apply left rudder pressure to stop the turn.

It's very important that you start the *timing* when you start the *pressure* to begin the turn, and start the roll-out *pressure* when the proper time has elapsed. There's no way to do this right without practice; with the AI and DG covered, run through the proficiency exercises in Chap. 21, with emphasis on developing consistency in your roll-in and roll-out timing.

The magnetic compass must be used *only* to check heading before or after a turn. There are too many opportunities for confusion and disorientation to make it useful in any other mode. If a controller requests a turn, or if you need to change heading on your own, *always* take a few seconds to check the reading on the mag compass while it's stabilized. And when you roll out, give the compass a few seconds to settle down before taking any rash measures to correct your heading.

When only small heading changes are required, press on the rudder pedal until the needle or airplane symbol has deflected one-third of the standard-rate indication and count the degrees of turn at 1 per second. Better to make two or three very small corrections and wind up where you want to be than one huge heading change that throws everything out of kilter.

# ✈ The Secret of Altitude Control

With no direct indication of airplane attitude, you'll find it rather difficult to establish consistent climbs and descents by trying to combine pitch changes and power settings. But take heart; there's a one-input way to change altitude, and it works every time.

Consider for a moment the aerodynamics of elevator trim. When your airplane is stabilized in level flight at a given airspeed, with all control-column pressure trimmed away and your hands off the wheel, the elevator is responsive only to changes in airspeed. If power is increased, the nose will pitch up to maintain that trim speed, and the result is a climb at or very close to the trim airspeed. Reduce power a bit, the nose will drop just enough to maintain the trim speed, and the airplane will enter a descent—all with no pilot input except the power change.

Most light airplanes will respond predictably to power changes. An increase of 5 inches of manifold pressure or 500 rpm will produce roughly a 500 fpm climb; reduce power the same amount, and you can count on approximately a 500 fpm descent. (Experiment with your airplane to arrive at more accurate power settings.) These are vertical speeds that will surely suffice for continuing the flight or getting down to approach altitude, and if you plan properly, 500 fpm will also do an acceptable job of descending to minimums on an instrument approach.

Single-engine propellor-driven airplanes are particularly susceptible to yaw effects when power is changed. Bear in mind that, when increasing power, you must counter the left yaw with pressure on the right rudder. How much? Just enough to keep the turn needle from moving—and remember that you'll need to *maintain* that right rudder pressure throughout the higher-power operation. The opposite will be true when power is reduced. Bottom line for all situations: If you don't want the airplane to turn, use whatever rudder pressure is required to prevent a heading change.

Some airplanes tend to climb and/or descend at airspeeds slightly different than trim speed, so once again, experiment with your airplane *before* the vacuum pump goes out or a gyro fails, and you'll know what to expect. In any event, it's hands-off-the-wheel time again. When an altitude change is required, do it with power; pitch will take care of itself.

# ✝ The Partial Panel Approach

If you were to survey instrument pilots, you'd probably find that most of them would prefer an ILS approach when operating on partial panel. Sound thinking because the ILS promises the best chance of finding the runway in bad weather, and when you're in trouble, you shouldn't turn down anything that will help you get on the ground safely.

An airport equipped with ILS will more than likely have a radar approach control facility, which means that you won't have to do all the navigating yourself. Be sure the approach controller knows of your plight, and you'll probably be provided a "no-gyro" approach in which "turn right/left" and "stop turn" commands replace the assignment of specific headings. Timed turns (3 degrees per second) are assumed in this case, and the controller will start/stop your turns to line you up with the final approach course for an instrument procedure or for a radar-controlled approach for one of the runways. Use the same steady, measured rates of roll-in and roll-out you've practiced, commencing rudder pressure on the controller's commands.

In the case of a nonprecision, no-radar approach, fly the procedure exactly the way it's published, with no shortcuts. Give yourself extra time on all the segments. This is no time to be rushed; do it right the first time.

Pilots of retractables have a distinct advantage at this point because the landing gear makes a very effective supplementary altitude control. In level flight with gear retracted, the amount of thrust is exactly that required to maintain altitude; the drag of extended wheels has the same effect as reducing power, and most light retractables will descend at about 500 fpm in this condition—again, no pitch inputs are required from the pilot. To level off with gear extended, it's necessary only to add enough power to make up for the gear drag. How much? Whatever is required to stop the altimeter from unwinding.

In the unlikely event of a missed approach, the pilot of a retractable need only raise the landing gear, thereby removing drag, and the airplane will enter a climb—shallow to be sure, but a climb—with no more action than the flip of a switch. You should of course add more power to reach a safe altitude in a timely manner, but the "gear-up" climb procedure gets you started with a minimum of effort and distraction. You'll also notice that, because there's no power change, there's no yaw induced when entering or terminating a descent with the landing gear.

## ✛ Easier Said Than Done

Reading all this in the comfort of your favorite chair, in a room that you *know* isn't going to spiral into the ground, the specter of partial panel seems completely manageable. In the real world, however, you'll find that a completely different set of circumstances exists. Loss of your primary attitude instruments in the clouds is going to be scary and filled with potential for a genuine catastrophe. Survival probably depends in great measure on your state of mind, and you can get rid of some of the anxiety by letting someone (probably the controller who's working your flight) know of your problem early on.

Preparation and practice are undeniably helpful in holding the partial panel monster at a respectable distance. Any of the proficiency exercises in Chap. 21 will provide ample opportunity for development of your partial panel skills. Don't be satisfied until your control of the airplane is *good,* and don't let up—stay proficient with frequent and regular practice. It's the *only* way to fly.

There were numerous instances of vacuum pump failures in Cessna 210s in the late 1970s and early 1980s. The subsequent partial panel experiences proved catastrophic for a number of pilots, including the one in the following example, who jumped from the frying pan into the fire.

The preflight briefing indicated a significant cold front oriented northeast to southwest, lying across the route of the northbound flight. However, tops were reported at 10,000–12,000 feet, and the pilot elected to climb over the clouds, out of the icing and turbulence.

Just a few minutes after leveling at 12,000 feet, the pilot announced, "I've lost vacuum. I'd like immediate clearance back to Knoxville" (the departure point). The controller complied right away, "Cleared direct Knoxville, descend and maintain 8000."

A minute later, the controller had coordinated a lower altitude and cleared the flight to 6000 feet at the pilot's discretion; the pilot indicated that he was on his way down, didn't need any special handling, and that "everything's okay except the vacuum." In less than 3 minutes, the C210 had spiraled into the ground, with no survivors. Spatial disorientation strikes again.

This pilot was acutely aware of his problem, and even though the airplane was probably on top when the vacuum failure was detected, he chose to descend through the clouds to get back to the departure airport.

Postaccident investigation revealed that clear weather existed only 50 miles to the east, ahead of the cold front. Why didn't the pilot stay above the clouds where he had a natural horizon? Why didn't he ask for more weather information, or at least proceed eastbound, where basic weather knowledge would suggest better conditions?

There are no answers to those questions, but if a similar situation ever shows up when you're flying, take a close look at *all* the options before you plunge into the clouds with less than a full instrument panel.

# 5

# IFR Flight Plans

It's nice to have the flexibility of filing or not filing a VFR flight plan, being able more or less to come and go as you please, yet having the protection of a flight plan when you want it. But when the weather goes bad and instrument flight is necessary, the rules change—there is no longer a choice for the pilot operating in controlled airspace. Full-blown or abbreviated, flight plans are a way of life for instrument pilots; they're an absolute necessity for IFR operations in controlled airspace. Use of the proper filing procedures and a full understanding of what happens when you file can save considerable time, make your IFR operations more efficient, and enable the system to serve you better.

There are several ways you can file a flight plan with a Flight Service Station—in person, by telephone, by computer, or by radio—but they share one commonality; the format is as standard as peas in a pod. Whether you file IFR three times a day or once a month, you can exhibit professionalism and save a lot of time by studying the flight plan format and using its sequence every time. This flight plan format (Fig. 5-1) is essentially the same one the FSS specialist calls up on a computer screen when you call and say you would like to file. Notice that each block contains a question to which you are going to supply the answer, so there is no need to repeat the name of the item. A properly filed flight plan should consist of a series of numbers, abbreviations, and names. No questions please, just the answers! Whenever you file and the specialist has to ask you to repeat something, you haven't done it right.

| DEPARTMENT OF TRANSPORTATION— FEDERAL AVIATION ADMINISTRATION | | | | | Form Approved OMB No. 04-R0072 | | |
|---|---|---|---|---|---|---|---|
| **FLIGHT PLAN** | | | | | | | |
| 1. TYPE — VFR ✓ IFR DVFR | 2. AIRCRAFT IDENTIFICATION *1234A* | 3. AIRCRAFT TYPE/ SPECIAL EQUIPMENT *BARNBURNER 408/A* | 4. TRUE AIRSPEED *170* KTS | 5. DEPARTURE POINT *BNA* | 6. DEPARTURE TIME — PROPOSED (Z) *1410* — ACTUAL (Z) | | 7. CRUISING ALTITUDE *7000* |
| 8. ROUTE OF FLIGHT | | | | | | | |
| *VSE GARVY X, V243 BWG 150/10* | | | | | | | |

| 9. DESTINATION (Name of airport and city) *BWG* | 10. EST. TIME ENROUTE HOURS *0* / MINUTES *20* | | 11. REMARKS *ROUTING REQUESTED DUE TO WX* | | | | |
|---|---|---|---|---|---|---|---|
| 12. FUEL ON BOARD HOURS *3* / MINUTES *30* | 13. ALTERNATE AIRPORT (S) *STANDIFOLD FIELD LOUISVILLE, KY.* | | 14. PILOT'S NAME, ADDRESS & TELEPHONE NUMBER & AIRCRAFT HOME BASE *J DOE ON FILE BNA FSS* | | | | 15. NUMBER ABOARD *5* |
| 16. COLOR OF AIRCRAFT *FUSCHIA, MAGENTA & CERISE W/BURGUNDY TRIM* | CLOSE VFR FLIGHT PLAN WITH_____FSS ON ARRIVAL | | | | | | |

FAA Form 7233-1 (5-72)

**Fig. 5-1** *Flight plan form, properly completed for the proposed flight from Nashville to Bowling Green.*

## ✢ Get Off to a Good Start

Assuming that you have received a weather briefing, decided on a route, and are filing with FSS by telephone, you can provide the information for the first two items right off the bat by saying slowly, "This is Barnburner 1234 Alpha with an IFR flight plan." When the FSS specialist rejoins with "go ahead," you may assume that your type of flight plan and aircraft number have been inserted in Blocks #1 and #2. (DVFR will be discussed later in this chapter.) You've blown the quest for efficiency if you repeat these items, so charge on, beginning with Block #3.

Aircraft Type deserves some emphasis: Cessna or Piper or Beech by themselves don't tell the story, so make your description adequate. Use model numbers for most airplanes, for example, Cessna 182, PA-28, BE-20. Everyone knows that a DC-3 is a Douglas, and nobody but Beechcraft has ever built a D-18, but who ever heard of a "Barnburner"? It would be prudent, therefore, to refer to your flying machine as a specific model of Barnburner.

In this same block, an indication of special navigation equipment is requested. This is vital information for Center controllers because it helps them with subsequent clearances, requests for position reports, and so on. Know which suffix applies to your airplane; if you fly several variously equipped aircraft, you might want to make a note of the suffix code:

/X   No transponder

/T   Transponder with no altitude
     encoding capability

/U   Transponder with altitude
     encoding capability

/D   DME, no transponder

/B   DME, transponder with no
     altitude encoding capability

/A   DME, transponder with
     altitude encoding capability

/C   RNAV, transponder with no
     altitude encoding capability

/R   RNAV, transponder with
     altitude encoding capability

/W   RNAV, no transponder

/G   GPS, with enroute, terminal,
     and GPS approach capability

Moving on to Block #4, True Airspeed, say only the numbers, such as "one seven zero." This figure may vary somewhat due to temperature changes aloft, but you should be able to hit it fairly accurately. Experience will provide a workable number for this block, but when in doubt, use indicated cruise airspeed plus 2 percent for each 1000 feet above sea level, and you'll not be far off.

Block #5 should be the airport's official three-character designator (such as IND, FAT, and 3G4) or the airport name if no designator has been assigned, and Block #6 should be a reasonable estimate of the time you expect to be ready for takeoff. Always given in Zulu, or Universal Coordinated Time, it should not be less than 30 minutes from right now, the time you are filing. It usually takes that long for your proposal to filter through the system from FSS to Center and back to FSS or Ground Control or Clearance Delivery, depending on the airport facilities. If you need to get airborne sooner, explain the situation and ask FSS to expedite your clearance. They'll do all they can to help, but it is ultimately up to Center to fit you into the traffic flow. The busier the terminal, the slower this process will be, but it's worth a try. Some very busy areas (Chicago, New York, Los Angeles, for example) would like to have your request at least an hour before departure. Since you're dealing with Zulu time, know what the conversion factor is and don't ask the specialist to do your arithmetic for you; time zones and conversions for each are found on IFR enroute charts.

If it appears you won't be able to make your proposed takeoff time, let ATC know about it because, as soon as your flight plan request comes into the center, tentative plans are made to accept it into the system. The computer will usually hold your request for an hour after the ETD (sometimes longer—check with FSS when you file), and you should amend this time if you can't possibly make it. It's simply a matter of calling FSS or Tower and asking them to revise your estimated time of departure. They'll appreciate your cooperation and can then let someone else use the airspace they were holding for you. It's the old "golden rule" trick.

A tragic accident several years ago was caused in part by a pilot's failure to update his ETD. He filed early in the evening and then suffered through the charter pilot's curse: passengers who always show up hours after they promised. They finally arrived in the wee hours, when the tower was closed. Rather than call FSS for a void time, the pilot elected to depart VFR and get his clearance in the air. That was fine as far as the weather was concerned (severe clear at takeoff time), but when the pilot called the departure controller, he was told that his clearance had "timed out"—that's ATC talk for "the computer has swallowed your clearance, give me a couple of minutes to find it." While the controller was hunting, the pilot continued on his way, which was unfortunately blocked by a mountain, and all aboard perished in the ensuing crash. Granted, the pilot should have avoided the mountain, but his IFR clearance would have contained departure instructions designed to keep him well clear of the rising terrain. When in doubt, find out how long your clearance will be retained. And when you know you'll be late, let ATC know.

Your altitude request (Block #7) is based on a number of factors, such as the hemispheric separation rules, wind, weather, and minimum enroute altitudes, and should be decided upon before filing. The key word here is "request" because an IFR flight will always be assigned an altitude by ATC. Don't argue about "maintain 9000" on a westbound flight; the controller has a good reason for it and will probably amend your clearance later on. If you're filing in the High Altitude Route Structure (18,000 feet and above), the same general rules apply until you reach FL 290 and above—altimeter errors can really build up at these levels, so 2000-foot separation is required. "East-odd, west-even" doesn't work anymore.

Block #8 can be either delightfully succinct or unnecessarily complicated, depending on your knowledge of what you can and cannot do when planning the route for your flight. (This chapter is concerned only with the language you should use when filing the flight plan; Chap. 6, "Preflight Planning," goes into detail about which airways departure routes, and so on to choose.) The route for any IFR flight, no matter how short, must be made up of three segments: departure, enroute, and arrival. You take off from one airport (departure) and navigate to some radio fix (enroute) from which you can execute an approach to another airport (arrival). If you keep this sequence in mind as you file, it will reduce the number of words required to communicate your request to ATC.

The departure phase can consist of a SID (Standard Instrument Departure), radar vectors, direct flight to a VOR, or immediate interception of an airway (if it's reasonably close to the departure airport). So your initial response for Block #8 should be a concise description of how you intend to get to your first fix; for example, "Boxer Two Departure to ALB" (identifier of the first VOR), or "radar vectors to ALB," or "direct ALB," or "Victor 14 to ALB."

If you harbor a suppressed desire to drive everybody in ATC right up the walls of their windowless control rooms, make it a habit to choose the most complicated, zig-zagging routes, using all the airways on the chart. Or make it easy on yourself and the system by picking out an established route, using wherever possible just one numbered airway to get where you're going. When you reel off the enroute portion of Block #8, use the Victor airway numbers and VOR identifiers.

More often than not, one airway will do the job. For example, referring to Fig. 5-2, consider a flight from Nashville, Tennessee (BNA), to Bowling Green, Kentucky (BWG). Victor 49 runs straight as a die between the two airports, so why not file that way? "Victor 49 BWG" is all you need to say about the departure and enroute segments. You've already mentioned the airport from which you're departing, and the next block is reserved for destination airport. The enroute portion should always wind up with the last radio fix you intend to use.

But suppose you're made aware of a line of thunderstorms along Victor 49, and you wisely decide to take an eastern detour. File "Victor 140 HARME, Victor 243 BWG."

In any event, make certain that the final entry in Block #8, Route of Flight, is an easily recognized navigational fix; in case you lose communications, the controllers will know exactly where you intend to go before commencing an approach to the airport. In Fig. 5-3 (the plan view of the VOR procedure for Bowling Green), it's obvious that when you're inbound on Victor 49 you are nearly lined up with the final approach course, and the center controller would probably issue a vector to intercept the 204 radial when you were close enough.

But notice that two IAFs (Initial Approach Fixes) are located conveniently 10 miles from the VOR on the airways leading to BWG from southwest and southeast. Should you be approaching on these airways, it would be sensible to end the Route of Flight with the appropriate IAF; for example, "V243 BWG 150/10," which tells the computer that you'd like to begin your approach from the BWG 150 radial, 10 miles out.

There are times when it's advantageous to file direct, off-airways routes, and your flight plan should consist of the first and last VORs you intend using, with the identifiers of the ones in between that will

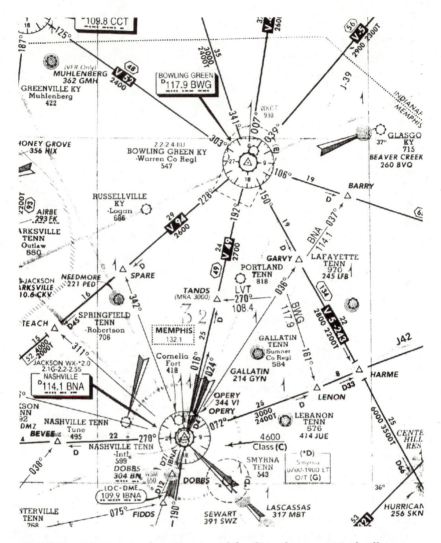

**Fig. 5-2** *Airways and navigational facilities between Nashville, Tennessee, and Bowling Green, Kentucky. (Copyright © Jeppesen Sanderson, Inc., 1979, 1997; all rights reserved; not to be used for navigation)*

serve as checkpoints; for example, "LVT, BWG, CCT, MWA." (The restrictions that apply to off-airway flights are discussed in Chap. 6.)

Having dealt with the departure and enroute phases of filing, turn your attention to the arrival portion. For planning purposes, the enroute portion of an IFR flight terminates over a convenient IAF or the navaid itself when IAFs are not designated on the chart; the published approach procedure will get you from there to the airport. Since the Bowling Green Airport is located only a couple of miles from the VOR, your complete route of flight is covered with "Victor 49 BWG." You'll receive an approach clearance before getting there, but in case your radios quit, ATC will know that you will fly to the VOR and commence an approach at the appropriate time. If Glasgow, Kentucky (just east of BWG), were your destination, it would be proper to file "Victor 49 BWG, direct GLW."

Destination, like point of departure, is adequately identified by the name of the airport unless there are several terminals in the area. "Bowling Green" would suffice here, since there are no other airports with which it might be confused. Whenever the aerodrome moniker is different from the city name, it's wise to include both in Block #9, so that Approach Control vectors you to the right airport—they'll assume you want to go to Big City Municipal unless you indicate some other airport as your destination. It's even wiser to always use the airport's three-letter identifier; that eliminates all the guesswork.

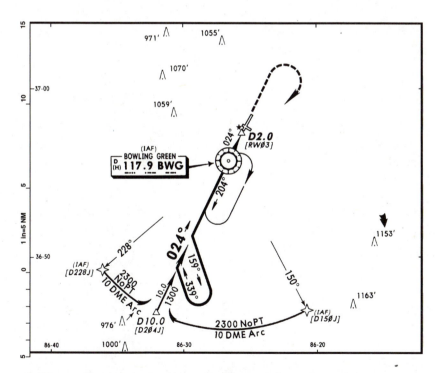

**Fig. 5-3** *Plan view of VOR approach procedure at Bowling Green, Kentucky. (Copyright Jeppesen Sanderson, Inc., 1985, 1996; all rights reserved; not to be used for navigation)*

The reason for Estimated Time Enroute (Block #10) is not to see how accurately you can preplan an aerial adventure; it's something to hang your hat on when the radios fail and you're wondering when to come down from your lofty, silent perch to begin an approach. In the absence of a more relevant time, the answer is your ETE as indicated in the flight plan plus your takeoff time. (See Chap. 12, "Communications Failure.") When filing the flight plan, it's sufficient to use hours and minutes to the nearest five—just the bare numbers, no further explanation needed.

Block #11 is a catchall, and rarely used; however, there are times when it can save you (and ATC) time and trouble. In the previous example (Nashville to Bowling Green), you elected to detour to the east because of weather. But this is not the obvious airways route between the two cities, so Center would probably clear you on Victor 5 *unless* you indicated in the Remarks block, "Requested routing due to weather." Or in parts of the west where higher altitudes are required, you may wish to impose a personal limit of 10,000 feet for physiological reasons; shortstop any haggling about the route you have chosen by inserting "no oxygen on board, unable higher than 10,000" in Block #10. This is also the proper place to insert "NO SIDs/STARs" if you haven't the appropriate publications on board. Flying into Canada and some other foreign countries or returning to the United States, you may insert the code word "ADCUS" (ADvise CUStoms) so they will be there when you arrive. Anything that bears directly on the operation of your flight can be stated in the Remarks block.

Block #12 (Fuel on Board) is a "nice to know" sort of thing for Flight Service, but it has no legal significance, since the pilot in command is responsible for having the proper amount of petrol in the tanks at the start of every IFR flight. (This rule is also discussed in Chap. 6, "Preflight Planning," and is a regulation which should never, never be broken, or even bent.) Having your endurance time on record can help Flight Service in a search and rescue situation, because it will give them at least a rough idea of how far you might have flown. This block should also be addressed with hours and minutes to the nearest five, such as "four plus four five," or "three plus two zero."

The selection of an alternate airport (if indeed, one is required at all) is a rather complicated process that is treated in depth in Chap. 6, "Preflight Planning"; the concern here is the proper way to express it to Flight Service, and Block #13 needs nothing more than the identifier (or name) of the airport. If it is something other than a self-explanatory name, include the city or town nearby.

Block #14 can have legal implications if there is more than one qualified pilot on board. The person whose name is placed here (last name and initials are sufficient) may be the one considered PIC should any legal action arise as a result of this flight.

Number Aboard (Block #15) is requested only for the customs people on international flights, and you are not really required to furnish this information for domestic trips. (Besides, there may be times when it's nobody's business how many people are on the airplane!) But to save a lot of grief for the rescue folks if they're needed, tell Flight Service how many souls are on board—just speak the number. "Souls on Board" is often abbreviated SOB, and there will be times when you list five SOBs on your airplane and *mean* it!

Block #16 should indicate the predominant aircraft colors, useful information should you alight at some location other than the one you had planned, such as on a mountain or in the desert without benefit of an airport.

All that's left to do is file the flight plan, so if you're using the telephone, the trip from Nashville to Bowling Green should be filed like this:

FSS: Good morning, Nashville Flight Service.

YOU: Good morning, this is Barnburner 1234 Alpha with an IFR flight plan.

FSS: [*After a slight pause to type the information in Blocks #1 and #2*] Go ahead with your flight plan.

YOU: [*Take it from Block #3*] Barnburner four zero eight slash alpha, one seven zero, Nashville International, one four one zero, seven thousand; Victor 140 HARME intersection, Victor 243 BWG one five zero slash one zero, Bowling Green, zero plus two zero, routing requested due to weather, three plus three zero; Standiford Field, Louisville [*or "no alternate" if that is the case*]; J. Doe, (your phone number and home base); white with red trim.

FSS: Roger, we'll put it on file; have a good trip.

If you have done your homework and filed the flight plan "by the numbers," the whole procedure will take about 1 minute; it's a definite timesaver for both parties, and you have invested yourself with an aura of professionalism.

## ☦ Same Thing, This Time by Radio

Filing a flight plan from the air usually results from one of two situations: You have encountered unexpected weather on a route that appeared VFR from the forecasts, or you want to have an IFR flight plan ready for the next leg of your trip. Of course, you can file a whole series of flight plans with the original Flight Service Station and they will pass your requests to the proper people along the route. But this approach is not always feasible, and it opens another door to the possibility of your flight plan being lost in the communications shuffle.

Center Controllers are not in the business of copying flight plan requests, so unless you're really in a bind, don't bother them; take your problem to Flight Service. Contact the nearest FSS and, when the specialist is ready, go through the names and numbers routine just as you did on the telephone. You should include the type of flight plan and your aircraft identification in the initial callup, such as "Broken Bow radio, Barnburner 1234 Alpha with an IFR flight plan." When the specialist says "ready to copy," start right off with Block #3. From here on, there's no difference in the filing procedure. FSS will tell you either to stand by for your clearance (if you need it shortly, the specialist will call the Center on a direct line and probably get it in a few minutes), or will tell you when, where, and whom to contact.

So much for getting a flight plan request into the system for a subsequent part of your trip; the other situation grows out of your need to obtain a clearance right now. Not an emergency, but there you are in absolutely beautiful VFR conditions without a flight plan of any kind, and you discover that your destination airport has just gone IFR. The same lag in communications, though not so extreme, exists when you contact Flight Service by radio, and again, "waiting" is the key word; why not contact Center directly, since you are only asking for clearance into the terminal area in order to make an approach? (Or if you can get into the terminal area VFR, call Approach Control.)

The first job is to find the proper frequency on which to call Center, and the enroute chart provides this information. Each ARTCC area of responsibility is subdivided geographically, and the chart shows the coverage of these sectors. The sector symbol (check the chart legend) will provide the proper frequency for that general area. Make your first call a short one in case Center is busy: "Memphis Center, Barnburner 1234 Alpha." When the controller answers, the following dialog might take place:

YOU: Memphis Center, Barnburner 1234 Alpha, 10 miles east of Walnut Ridge, 8500, squawking 1200, requesting an IFR clearance to Walnut Ridge Municipal.

CENTER: Roger, 34 Alpha, squawk ident. (Short pause, then—) Barnburner 1234 Alpha, you're in radar contact, cleared direct to the

Walnut Ridge VOR, descend to and maintain 6000. Say your aircraft type and true airspeed.

YOU: Barnburner 34 Alpha is cleared direct to the Walnut Ridge VOR at 6000; it's a Barnburner four zero eight, true airspeed one seven zero.

CENTER: 34 Alpha, roger; expect the VOR Runway 27 approach at Walnut Ridge.

That's all there is to it; you're in the system and can now proceed just as if you had been IFR all the way. A very efficient shortcut to the full-blown filing process, this method is much faster than going through Flight Service, although it is somewhat less preferable to filing IFR from the beginning. Don't expect this kind of service in one of the high-density terminal areas like Chicago or New York—they just don't have the time—but in less busy parts of the country, your request will almost always be granted.

## ✝ IFR Flight Plan to a Non-IFR Airport

There are literally thousands of small airports around the country which are used every day by general aviation pilots without the benefit of an instrument approach. Some of them lie in uncontrolled airspace and are not subject to the IFR rules. But there are many in controlled airspace to which operations are conducted quite legally when the weather prevents VFR flight all the way. If there is an airport close by with a published procedure, you may be able to make the approach there, break it off when you arrive in VFR conditions, and proceed to your "little airport" destination.

This is a good plan, but only under certain conditions; you must be sure that the weather will permit a safe VFR operation from the end of the published approach to the non-IFR airport, the approach used to get you down to VFR conditions should not lead to a "No Special VFR" airport if the weather is pushing visual minimums, and you should be prepared to land at the IFR airport if you can't continue VFR.

Every bit as important as the other considerations is your responsibility to communicate to ATC your intentions as early in the flight as possible, preferably in the Remarks block of your flight plan request ("will proceed VFR to Little Airport"). You'll earn the undying enmity of the controllers if you wait until you're on final approach and fitted neatly into the landing sequence to let them know that you are going to "break it off" and go to some other airport. There is usually no problem if you will just make your desires known early in the game. It's a completely legal procedure and can work to your advantage under the proper conditions; it appears underhanded and less than professional when a pilot tries to "sneak" an approach for the purpose of proceeding to a non-IFR field. Unless your flight presents a traffic hazard for them, ATC will go along with you, sometimes even supplying vectors to the "little" airport.

# ✝ IFR Flight Plans from a "No-Facility" Airport

Well, here you are, back at the airport after a day's work with your clients in West Snowshoe, Montana. When you arrived this morning, the sun was shining in a clear blue sky and the forecast promised the same thing all day. Now, preflighting the Barnburner in cold rain falling from a 500-foot overcast, you're not so sure. Besides that, the FBO has gone home—thank the gods who look out for pilots that there's a telephone booth outside the hangar.

All is not lost because there is still a way to obtain a clearance under these conditions; the only requirement is that you are able to contact someone in ATC: a Flight Service Station, a nearby tower, or even Center. (If there's plenty of ceiling and visibility, you might take off and remain VFR while you file your flight plan and receive clearance, but there's no guarantee of a quick clearance, and you might wind up flying around a lot longer than you'd like.)

A more efficient way to get the job done is to call the nearest FSS on the phone, explain the situation, and file a flight plan, with the additional request for a "void-time" clearance. (Choose the route and altitude to the first fix with great care because you will be entirely on your own until ATC acquires you on radar. File so as to get on an airway as soon as you can to take advantage of the guaranteed terrain and obstacle clearance. It may be necessary to refer to a sectional chart to make sure of the elevations along your route to the first fix.) Under these conditions, Flight Service will likely ask you to stand by, and will call your flight plan request directly to the governing Center while you wait. Bear in mind that this procedure is almost always undertaken from an out-of-the-way airport, and there will probably be little if any traffic in the immediate area. When a delay is probable, Flight Service will have you call back in a few minutes or will take your phone number and call *you* back; the latter is especially important when you have no more quarters.

If the system can handle your flight now, your clearance will be issued with a release time and a void time. You may not depart before the release time nor after the void time. The deadline imposed by the void time may not be very far away, but if it's too close for comfort, tell the FSS specialist and it will be revised. In anticipation of this, figure how long it will take you to board your passengers and taxi to the runway before you call for a void-time clearance.

It's important to get ready for a void time 10 or 15 minutes from now because Center doesn't want to hold airspace open any longer than necessary. If ATC hasn't heard from you shortly after the void time, they assume that you have taken off and suffered communications failure, and therefore, they must open up routes and altitudes for your entire proposed trip, just in case. This is one of those situations in which good pilots can exercise their judgment and refuse a void time that is too close. Remember that you will probably be taking off into IFR conditions (otherwise you wouldn't be concerned about the void time), and this is no time to be racing the clock. When you file your flight plan with FSS, let them know that, because of the distance to the runway or whatever, you cannot possibly be off the ground in fewer than 20 minutes (or a reasonable length of time according to the situation). FSS will pass this on to Center, and they'll more than likely respect your problem. From the time you enter the clouds, this type of clearance is no different than any other, and all the rules of IFR operations apply.

You can help yourself by anticipating the need for a void-time clearance. Suppose that during the afternoon in West Snowshoe you noticed the clouds beginning to thicken. If you suspect IFR conditions for your takeoff in the evening, call Flight Service and file your flight plan in advance, with the expectation of needing a void time. The FSS specialist may even be one step ahead of you and have a clearance waiting when you call.

Having received a void-time clearance, you must respect its limitations as if it meant life or death, which it might. Be sure your watch agrees with ATC time, and do all within your power to leave the ground not 1 second before or after the appropriate times. If you can't make the void time (and fuel-injected engines are notorious for not starting in a situation like this), shut everything down, swear a lot, run through the rain to the phone booth, call Flight Service, and start all over again. It might be wise to toss a couple of quarters into the glove box of your airplane the next time you fly; you never can tell when your passengers might be as broke as you are!

## ✝ Restricted Airspace

When it comes to flying through restricted airspace, happiness is having the word "instrument" on your pilot certificate. As long as you plan your flight on the Victor airways, you can rest assured that ATC will not let you traverse any airspace which is also being used by artillery shells, rockets, or the annual meeting of the U.S. High-Altitude Kite Flying Association. When you ask for a direct, off-airways routing that will cross through restricted airspace, your request will not be honored unless and until the ATC facility with which you are working has coordinated with the using agency and made certain that it will be safe for you to pass. Whether you're operating on a clearance or not, be extremely alert when in or near a Military Operations Area (MOA); especially when they're going about the business of training aerial warriors, the high speed and small profile of training aircraft make them very difficult to see.

A *prohibited area* is the most restrictive of all the limited-use airspace, and its label means exactly what it says. If it appears that an IFR clearance will take you through such airspace, it would be wise to query the controller and get the monkey off your back. (When you're on a VFR flight, trespass of a prohibited area might get you an audience with somebody in the FAA, maybe even the FBI.)

## ✝ IFR and ADIZs

National security is the only purpose for establishing ADIZs (Air Defense Identification Zones), and they are primarily concerned with aircraft approaching our shores and borders from the outside in. There is a part of the Federal Aviation Regulations set aside to describe this airspace and to lay down the laws for operating therein. Again, happiness is having an instrument rating because your ATC clearance into or through an ADIZ takes care of all the reporting and filing requirements. There is a flight plan option—DVFR, Defense Visual Flight Rules—for pilots who wish to fly through an ADIZ without filing IFR, but once you have an instrument rating, why not use it, whether there's a cloud in the sky or not? Of course, if you happen to be an aerial photography buff and would like to get some closeup shots of our Air Force's finest interceptors, fly out to sea 100 miles or so, then turn around, and head for shore, fast. You'll get your pictures, they'll get theirs, and in addition, you will be given the opportunity for a chat with representatives of several government agencies!

The ADIZ rules do not apply to certain operations depending on true airspeed and route of flight (check the *Aeronautical Information Manual* for the current restrictions). But the same philosophy that kept you out of trouble in other types of restricted areas will work just as well here; when in doubt, file IFR, get a clearance, and you're home free. ATC knows where you are, who you are, where you're going to and coming from, and when this information is passed along to the Air Defense Command people, everybody's happy.

# 6

# Preflight Planning

It was spelled out in letters a foot high, above the door to the flight line at a military pilot training base: PLAN YOUR FLIGHT, AND FLY YOUR PLAN!—a good piece of advice for any pilot. When you're flying IFR, there will be diversions and changes and nonstandard procedures, and on occasion, ATC may require you to go rather far outside the routing or altitude you had planned for the flight, but having a plan puts you in the driver's seat right from the start. When changes come up, you have at your fingertips the information you need to make a decision: go or no-go, accept a clearance or reject it, try an approach or go directly to your alternate. You're asking for trouble when you take off knowing nothing of the enroute weather or wind, or where you can go if everything turns sour.

## ✛ A Preflight Checklist

No one is going to double-check your preflight planning, so it's up to you to do it right every time. The easiest and surest way to accomplish this is to do it the same way every time by using some kind of checklist or at least a routine information-gathering and decision-making process that leaves nothing out of the planning picture. Looking at the situation through real-world glasses, there will be many short IFR trips that don't require in-depth preparation of the sort you should go through for an extended aerial journey, especially into strange territory. However, when you form good habits, you will tend to include the critical items even though you're planning a short, uncomplicated flight.

## ✚ Charts and Equipment

Like a mechanic starting a 100-hour inspection without a toolbox, you
will be behind the power curve right away if you try to launch the planning process without the proper equipment. It's true that weather will
be the most significant factor in planning, but how can you consider alternative routes around the weather or over more hospitable terrain unless you have the charts handy? The scope of your operations and the
type of flying you intend to do determine how many charts you carry
around. For basic IFR work, all you need are the appropriate enroute,
area, and approach and landing charts. SIDs and STARs are nice to
have, especially if you frequently fly into and out of airports where they
are used; they will save you and the controllers a great deal of time.

The equipment that you have assembled to carry with you is a matter as personal as the clothes you wear. Some pilots carry more gear in
a Cessna 172 on a 30-minute flight than a 747 captain needs between
New York and Paris. Whatever you settle on as minimum equipment,
spend a few dollars for a flight bag of some kind to keep things organized; something that will hold what you need in flight, yet not inter-
fere with seating or control movement. There are a number of specially
designed cases on the market that will hold a couple of approach plate
binders, your enroute charts, two ham sandwiches, and a paperback
novel. There are also cases available that rival steamer trunks for
capacity; you have everything you'll ever need, plus a lot of stuff you'll
*never* need. Leave these to the big-airplane pilots, who really need all
that paperwork.

In addition to the charts, you should have some kind of a computer with which to keep track of your time/distance and fuel situation. As your experience grows, you will be able to estimate time
enroute very closely by applying an average wind component to the
true airspeed, but for the sake of accuracy, use a computer—
especially when the proposed flight is long enough or the headwinds
strong enough to generate concern about having sufficient fuel to get
where you intend to go. By all means, include a good strong flashlight and a couple of spare batteries; you never know when you
might be caught out after dark, and a pack of matches or a cigarette
lighter just won't do the job if the electricity quits.

## ✚ Checking the Weather

Without a doubt, the biggest single factor affecting your planning for
an IFR flight is the weather, but you can drive yourself right up the

wall worrying about it too far in advance of your trip. At 48 hours before takeoff time, the very best the weather people can do is give you a general idea of weather conditions; at 24 hours, the forecasts become more accurate, but in most cases, you can't really obtain a good picture until just before takeoff. (Of course, this is weather information for *planning* purposes; you should always be checking a tight weather situation right up until you walk out to the airplane and continue checking it enroute.) Always be thinking of some alternative course of action in case the weather gets worse than you can put up with. If the forecasts are filled with doom and gloom, reserve an airline seat, plan to drive, or make arrangements to go another day.

When it looks as if the weather is going to put you between a rock and a hard place, get out the instrument charts to determine whether you can make it under the freezing level with respect to the minimum IFR altitudes or whether there is a convenient detour available. Do you have the range to make it nonstop? Does the destination airport have an approach good enough for the weather that is forecast? Don't take it for granted that all ILS approaches will bring you down to 200 feet and a half mile; there are a number of them that have much higher minimums because of terrain, and the same philosophy applies to all types of instrument approaches, so check and be sure.

Once you have decided you can get from here to there in one piece, there is another source you should consult to prevent an awkward, perhaps expensive, situation. It is considered the last word in operational information (short of being there, of course) and bears the name NOTAM—governmentese for NOtice To AirMen. Whenever a radio aid goes out of service or airport lighting systems fail or parachute jumping is in progress (how'd you like to have a group of four parachutists in formation suddenly join you in the cockpit of a Cherokee?) or any one of a thousand things that could affect the operation of airplanes, it will be published as a NOTAM in the ATC communication system. It would be embarrassing indeed should you take off for Buffalo in the middle of winter with an airplane full of passengers and find out halfway there that snowplows are the only vehicles permitted on the runways at Buffalo today. You can avoid this kind of problem by asking the briefer to check the NOTAMs while you're still in the planning stage.

You should be aware that there are several different types of NOTAMs, and even though you ask, you may not get all the information you need. The sleeper is the Class II NOTAM, which appears in printed form for mail distribution and which doesn't show up in the course of a normal scan. When you request NOTAMs during a preflight briefing, the specialist looks at the end of the hourly sequence for your destination or refers to the NOTAM summary. Unfortunately, the Class II NOTAMs don't show up on the screen, and you'll never know unless you specifically request, "Are there any Class II NOTAMs in effect?"

An accident at a midwestern airport several years ago points up the importance of checking *all* the NOTAMs. Following an uneventful flight from Chicago, the light twin commenced an ILS approach and was tracked by radar to the vicinity of the outer marker. Shortly thereafter, the airplane broke out of the low overcast in a vertical bank, struck several power lines along the main street of a small town, and crashed into a bookstore.

Pilot disorientation? Engine failure? Instrument malfunction? We'll never know for sure because all aboard were killed, and the airplane was completely destroyed by fire.

But an intriguing report was filed by another pilot who flew the same ILS approach an hour after the accident. He wrote, "I was making an ILS coupled approach when the plane abruptly turned right and assumed a nose-down attitude. The descent rate was 2000 fpm, and the heading changed approximately 130 degrees. I uncoupled the autopilot and hand-flew the plane to a level attitude, after losing 1000 feet."

Now the Class II NOTAMs become significant because right there on the ILS approach chart was the following limitation: *Autopilot Coupled Approach Not Authorized*. The same information was included in the Class II NOTAMs for that airport.

This is the sort of *very*-nice-to-know information that is published as a Class II NOTAM. There's no evidence that the pilot of the twin requested this information or that he noticed the restriction on the approach chart. For that matter, there's no evidence that a glitch in the ILS signal caused the accident. But the FARs require that instrument pilots become familiar with all information pertinent to a proposed flight. When in doubt, which should be frequently in this business, *ask*.

# ✝ Route and Altitude

At this point, you should have enough weather information to make a choice of route and altitude. Unless a detour is indicated because of weather, high terrain, or restricted airspace, you should be looking for the route that will get you where you're going in the shortest possible time. This usually means the nearest thing to a straight line between departure and destination. For long trips, you should consult an IFR planning chart, which shows the airways from coast to coast and border to border. (The Jeppesen service includes such a chart, or you can purchase one from the government.) You'll generally find an airway nearly parallel to the direct route. Unless you look at the big picture, you may miss the airway that goes all the way from point A to point B; filing one airway is so much easier than a bunch of short segments, and it eases the pain for ATC, too.

Preferred routes, with major terminals and the airways that ATC would like you to use when traveling certain city pairs, are listed in the Avigation section of the Jepp service and in the *AIM*. Preferred routes are the fastest, most efficient way to go. Although not always the *shortest,* they are ultimately the *fastest* because that's the way ATC is going to clear you, so you are better off filing the preferred route at the outset. (See Chap. 5, IFR Flight Plans, for the proper way to file a devious route because of weather or limiting altitudes.)

ATC's preferred routes can also be used to advantage when you plan a trip to a point short of the destination terminal mentioned in the route listing. If you use the preferred route as far as you can, you'll get better service from ATC.

Standard Instrument Departures (SIDs) are really short-range preferred routes limited to the departure phase of an IFR trip, and if you frequently fly from an airport with published SIDs, you'll find it worthwhile to have these charts in your flight kit. When SIDs are in use, plan your flight accordingly by noticing the point at which the SID puts you into the enroute structure. Your enroute planning should begin at the VOR (or other fix) specified as the end of the SID and use the preferred route, if applicable, from that point to destination.

At an increasing number of airports, STARs are coming into use. Not heavenly bodies to be wished upon or used for celestial navigation, these STARs are the acronymic way of saying Standard Terminal Arrival Routes. They are just SIDs in reverse and provide specific routings into a terminal area. When you are flight planning to an airport with a published STAR (big airports may have several), end the airways portion of your route at the VOR or intersection that is the entry, or gate, to the STAR procedure. Using the STAR when you file your flight plan will save everybody time and trouble because you won't have to be given a change of routing along the way. STARs are near-exclusive features of high-speed, high-altitude operations.

The ultimate in preflight planning, at least from a routing standpoint, would see you departing under the guidance of a SID, proceeding on your way via a preferred route, and winding up the flight with a Standard Terminal Arrival Route (STAR). The key is to realize that the air route structure is a huge system, and where the designers have identified channels (preferred routes, SIDs, and STARs) in the system, you are much better off putting yourself there as soon as possible after takeoff, and as early as possible as you approach the terminal area. (If you don't carry SIDs or STARs, don't forget to mention that fact in the Remarks block of your flight plan. It's the best way to let ATC know of your limitations.)

When a SID isn't published, plan your departure by the most direct route that will get you to a VOR or an airway as soon as possible after takeoff. The outstanding advantage of choosing an airway, direct route, or VOR radial to start your flight is that of definition; you know exactly which way to go in the absence of ATC instructions or if your radios quit before you can receive further clearance.

Radar control on departure is the name of the game in IFR flying today. At most terminals, you will be picked up on radar almost immediately after takeoff, and the controller will vector you around other traffic and on course just as expeditiously as possible. If you're not at all sure of the best way to get from the airport into the enroute structure, there's one more way you can get help; request radar vectors to the first VOR on your route, and Departure Control will lead you by the hand.

There are parts of the country that just don't have any airways headed in the direction you want to go or don't connect the particular points A and B you had in mind. So why not devise your own route, direct from one VOR to the next on your intended line of flight? There are no rules against flying direct (off airways), but there are several very important limitations you must consider before undertaking such a course of action.

1. You should not file a direct IFR flight between VORs that are more than 80 miles apart (this is a low-altitude limit that is expanded to 260 miles for flights at 18,000 feet and above). Designed for frequency protection, the 80-mile minimum guarantees that you will not pick up a signal from another VOR station on the same frequency.

2. You must choose a flight altitude that will provide the same obstruction clearance that is automatically taken care of by airway MEAs and MOCAs. You may have to consult sectional or WAC charts or other topographic references to be sure that your off-airways route is high enough.

3. Your course must be direct—in a straight line between VORs—no doglegs on this trip!

4. You'll not receive an IFR clearance for a route that goes beyond or below ATC's radar coverage.

Filing direct takes on new and wondrous significance when you're flying an airplane equipped with approved Loran or GPS. Now you can file from airport to airport and expect to be cleared that way. There are times when traffic volume or other ATC problems won't permit you to fly on such a long leash, but it's always worth a try. And if the first controller you talk to comes up with an airways clearance, ask again to go direct when you're handed off. Loran or GPS direct works nine times out of ten. (See Chap. 10, Area Navigation, for more details.)

In summary, you should consider the following order of routing priorities when planning an IFR flight (disregarding deviations for weather or other limitations and the availability of Loran or GPS):

1. The preferred route between departure and destination.
2. Victor airways that work out closest to a straight line between here and there.
3. A direct, off-airways route at an altitude high enough to provide comfortable clearance from terrain and obstructions (1000 feet above everything 4 miles either side of your course in the flatlands, 2000 feet in those parts of the country that qualify as mountainous).

# ✛ Altitude

Many pilots pay surprisingly little attention to the selection of an IFR altitude, even though it plays a huge part in putting the most miles behind you for the fewest dollars or making the difference between a miserable experience and an aesthetic adventure. Of course, there are some practical limits, like the Minimum Enroute Altitude (MEA), availability of oxygen, and the optimum performance altitude for your plane; turbulence, icing, wind, and weather problems cannot be ignored, but they'll be thoroughly hashed out later.

MEAs and optimum performance altitude provide the absolute floor and probable ceiling for a nonsupercharged airplane. You can't plan an IFR flight at less than MEA, and unless there's a real whopper of a tailwind, you will probably lose money by climbing above your airplane's optimum altitude. (This is usually the highest altitude at which the engine can maintain 65 percent of its rated horsepower.) Check the route for the highest MEA—that will eventually become your cruise altitude. Although you can file (and be cleared at) a lower altitude before you get to that particular airway segment, ATC will move you up to the MEA at the appropriate time, so plan on it. The abnormally high MEAs will occur over the mountains, because they must provide 2000 feet of obstacle clearance in those areas.

For normally aspirated (nonsupercharged) engines, it sometimes comes down to a choice between less performance at high altitude or a much longer route to avoid the higher levels. There's almost always a westerly wind at altitude, so clawing your way up to 12,000 or 13,000 feet in an airplane with an optimum altitude of 7000 is not necessarily a bad move if you're eastbound; the extra groundspeed on a long trip may more than make up for the time spent in the climb. There's only one way to decide, and that's to dig out a computer and compare the estimated time enroute for the high altitude versus a lower one. (A friend was proceeding eastbound in his nonturbo Bonanza a while back, taking full advantage of the tailwind at FL210. An airline captain overheard one of his position reports and inquired, "What's a Bonanza doing at 21,000 feet?" My friend answered with aplomb, "Struggling!")

In general, when weather conditions are not a factor, it is usually better to file as high as practical, at or above the appropriate MEA, and at or above your airplane's optimum altitude. The benefits that accrue to high flight often overcome the extra 5 or 10 minutes spent climbing, and maybe it will put you above an icing level, keep you out of turbulence, or treat you to groundspeeds you can brag about at the bar. And if nothing else, why not climb up above the choking brown gunk that is the rule rather than the exception over our country these days? Why not climb up into the only place left where you can see forever and, at least for a little while, treat your lungs to some really clean air?

There's also something to be said for the safety aspects of flying high. The very fact that fewer pilots use the higher levels means that there will be fewer people to run into at altitude; and given the usual situation of being in the clear (at least between cloud layers) when you're operating in the upper reaches of the sky, you'll have more time to observe and react to other air traffic.

At any rate, give "high flight" a try on your next few trips and decide for yourself. Keep records of how much time is required to climb, how much fuel is burned, and the difference in elapsed times for similar wind and weather conditions at lower altitudes. Then compare the hard facts with your memories of the smooth, crystal-clear air up there; remember the pleasant glow you experienced when that first-time passenger became a believer after flying high with you.

# ✝ Distance and Time Enroute

Now that you're settled on the route and altitude, add up the distance from departure point to the initial approach fix or the radio aid serving the destination airport. If you are not in the habit of computing time enroute, or whenever your aerial speed is an unfamiliar one, take the time to figure it out rather precisely. You'll be loading the situation in your favor, but perhaps more important, you will be practicing so that as your experience grows you will be able to estimate the time very accurately with a minimum of paperwork. The basis of the computation is true airspeed, plus or minus a wind component, arriving at an average groundspeed. All three of these speeds will change during the climb, so you will have to add a time factor.

For most light, nonturbocharged airplanes, a climb of less than 10,000 feet in still air will add about 5 minutes to the time it would take if you were at cruise airspeed all the way. You can figure it out for your airplane, but better yet, make a note of the time penalty for climbing to altitude on your next several trips, and you'll soon have a rule of thumb that will stand you in good stead.

Remember that you are asked to estimate your time enroute for only two reasons: to ensure that you'll have enough fuel on board and to provide you and ATC with a descent time should the radios fail. Nobody really cares how close your estimate is on a routine trip. One thing you don't want to be is ultraconservative and estimate too much time for the journey; if the radios *do* fail, you're going to hold over the destination fix until that time elapses, and if you throw in a half hour "just to be sure," you may regret it on the other end. It would be no fun going round and round in a holding pattern for 30 minutes in the clouds without radios, waiting for that ETA to come up—your nerves would be tied in double knots! Be as accurate as you can, but don't spend hours working out the ETE; it's just not worth that much of your time.

# ✝ Fuel Required

Disregard for fuel requirements is an open invitation for big trouble. There's no other situation in aviation that leaves you so frustrated, so helpless, and so upset with yourself as when you realize that you haven't enough gas in the tanks to get where you're going. It's bad enough VFR, but the pressure really builds when you're on solid instruments, and it will rapidly bring you to the PMP (point of maximum pucker). The most sensible way to keep from painting yourself into a gasoline corner is to observe an inviolate *1-hour* reserve fuel requirement on *all* flights. When it appears that headwinds, holding, icing, carburetor heat, traffic delays, or route deviations are eating slowly but surely into your spare gasoline, *do something,* and do it *now*. Landing at the nearest suitable airport is the recommended procedure.

I'll share with you the experience that made me a believer in the 1-hour rule. Before I had enough rank to say "no" to anyone wearing eagles on their shoulders, I agreed to undertake a nonstop flight from an airbase near Tokyo to our home field on the west coast of Korea. It was the middle of winter, solid IMC all the way, and of course, the headwinds increased, the icing got worse, and the fuel gauges in our C-47 (DC-3) sank lower and lower. In short, we overflew two perfectly good airbases where we could have refueled, finally broke out of the clouds on final approach at home, and on touchdown the right engine quit. Out of the 800 gallons in the tanks when we left Japan, there was a grand total of 28 gallons of gas remaining. Ever since, this pilot finds a place to land when the fuel supply gets down to 1 hour.

The fuel tanks in most light aircraft are large enough to sustain flight for 4 or 5 hours at commonly used power settings—more time than most of us care to sit in a confined space, to say nothing of the physiological problems that arise. It's much more comfortable to compute a reasonable maximum "time in the tanks" and vow never to fly more than 75 percent of that time. In a "4-hour" airplane, plan a pit stop at the end of 3 hours. Your passengers will love you for it, and you'll never come close to running out of gas.

Whenever you are operating in IFR conditions, you must have on board enough fuel to travel from departure to destination and have 45 minutes of fuel remaining based on the normal cruise fuel consumption of your airplane. Regardless of the fact that you will certainly reduce power whenever you are in a holding pattern, the 45-minute requirement at *cruise power* still exists, giving you a little additional cushion in a really tight situation. But this regulation, like all the FARs, is based on a *minimum standard,* so compute the fuel consumption for taxi, takeoff, climb, and cruise at the power settings you intend to use, subtract that total from the capacity of the airplane tanks, and if the resulting number doesn't correspond to at least 1 hour at the cruise power setting, you will be foolish to start out on that flight. The alternatives? Plan a fuel stop or wait for a day when the winds are a little less aggressive.

Here's a way to provide a solid planning figure with a little experimentation on your next short trip. Fuel system and safety considerations permitting, start up, taxi, take off, and climb to the highest altitude you normally use; do all this on one fuel tank (or set of tanks). When you level off, switch to a full tank and complete the flight. After landing, note the number of gallons required to top off the first tank. This figure plus the normal cruise fuel consumption rate will produce an accurate yardstick for future flights to make sure you have enough petrol.

## ✝ Alternate Requirements

It's perhaps one of the most abused regulations in the IFR book, but treating the alternate rule lightly can get you into serious trouble. Even though Flight Service Station briefers are quite aware of the weather conditions at your destination, seldom if ever will they remind you of the need for an alternate when you are filing a flight plan; the responsibility rests on the shoulders of the pilot in command.

The regulation that covers IFR alternates is cumbersome, wordy, and tends to drive pilots away from the consideration it deserves. There is a good rule of thumb that will *usually* do the job for you, and that is to get in the habit of filing an alternate for *every* IFR flight—pick the nearest airport with equal or better approach facilities within comfortable fuel range. Don't count on this procedure as an IFR panacea, but if everything goes down the drain, you'll have someplace to go. The worst situation imaginable would see you arriving at destination in weather that prohibits a landing, with no alternate filed, followed by a radio communications failure: You don't know *where* to go, ATC doesn't have any idea where you *might* go, and the resulting traffic hysteria would certainly get you an audience with an FAA inspector, if you should somehow manage to wander around through the clouds and get down safely without hitting anybody. You owe as much to the other pilots up there in the soup as you owe yourself, so always have an alternate airport planned before you climb aboard.

Like many of the rules that govern aviation, the IFR alternate regulation is restrictive in nature, letting you know what you may *not* do; but it can be reduced to convenient operational terms by asking yourself two questions before every instrument flight:

1. Must I file an alternate airport?
2. What airport may I use for an alternate when one is required?

# Must I File an Alternate Airport?

The answer to the first question depends on both the approach facilities available at destination and the forecast weather conditions for that terminal. If, as is usually the case, your destination has a published instrument approach, you can go on to the weather situation, which will tell you whether or not you need an alternate. But if the destination airport has no approved procedure, stop right there. An alternate airport is automatically required.

In an effort to keep you out of trouble, the government has decreed that either one of two weather conditions at destination demands an alternate be filed in your flight plan: One concerns the ceiling, the other refers to visibility, and they both respect a 2-hour time frame, from 1 hour *before* you expect to get there until 1 hour *afterward*. Ceiling and visibility would not normally change that much for the worse during a 2-hour period, and if this kind of rapid change were to take place, it would probably be adequately forecast or reported.

Here are the weather limitations about which you should be concerned: The reported or forecast *ceiling* (the lowest *broken* or *overcast* cloud layer) must be at least 2000 feet above the airport elevation, and the *visibility* must be at least 3 miles. (There's a way to remember it that's as easy as 1-2-3; 1 hour before or after your ETA, 2 thousand feet of ceiling, 3 miles visibility.) Table 6-1 presents the basic criteria you must consider when answering the first question: Must I file an alternate?

**Table 6-1.** Conditions and Answers to Question 1 Regarding the Need to File an Alternate Airport

| Condition | Must I file? |
|---|---|
| Destination has no approved approach procedure | Yes |
| Approach chart published for destination airport | No, unless weather is reported or forecast below specified minimums |
| Ceiling forecast or reported less than 2000 feet above airport elevation at any time during ETA ±1 hour | Yes |
| Visibility forecast or reported less than 3 miles at any time during ETA ±1 hour | Yes |

On occasion, you will encounter a weather situation in which the 2-hour time period contains "a chance of" or "intermittent conditions" or "brief periods of" ceiling and visibility that require an alternate on your flight plan. This problem has been hassled in the courts with little definitive guidance, so the sensible answer is to file an alternate whenever there's a question about whether it will be needed. It's so much better to have a safe haven in your hip pocket than to be forced to find one at the last minute. And how about the day when the forecast called for "clear and a million" when you filed, but now it looks like the sky has fallen at destination? That's easy: File an alternate with whomever you're talking when you discover that the bridges are burning ahead of you. Alternates aren't normally stored in center computers, but when you tell a controller, "I'd like to add Quincy as my alternate," the message will get through.

Notice that a forecast below *either one* of the conditions laid down by law (ceiling *or* visibility) requires an alternate. The complexity of the rule makes it even more sensible for you to form the habit of always filing an alternate for an IFR flight. You'll have the comfortable feeling that comes with knowing all the bases are covered, and selecting an alternate automatically will keep you from forgetting it on those rare occasions when it's needed.

The 1-2-3 rule will force you to plan for an alternate and carry the requisite fuel, but there are situations in which forecast weather conditions might meet the requirements of the FARs and still leave you out in left field. These situations are unique to high-altitude airports, and you should be aware of the potential problem when you're planning a flight into the mountains.

There are at least 10 airports in the United States where all of the MDAs are greater than 2000 feet AGL and/or the minimum visibilities are greater than 3 miles. When operating IFR to such airports, it is possible to encounter a situation in which you wouldn't *legally* require an alternate (and therefore perhaps not carry enough fuel to get somewhere else) and find on arrival that conditions are below published minimums. *Now* what do you do?

In addition to the airports just mentioned, there are others in the mountains that have MDAs only 100 to 300 feet below the alternate requirements. Again, only a slight worsening of weather conditions at your ETA could cause a problem.

In general, pilots should always be more diligent when planning a flight into mountainous areas (density altitude, capricious weather, rock-filled clouds, survival aspects, and so on), and when planning to fly IFR, that diligence should extend to a *guarantee* of enough fuel on board to fly to an alternate…just in case.

# What Airport May I Use as an Alternate?

This one is not nearly so difficult to answer (it shouldn't be, given the built-in conservatism of the first question), and it lends itself to a simple rule: If, at your ETA, the airport you wish to use for an alternate is forecasting a ceiling of more than 800 feet and a visibility of more than 2 miles, it's okay to go ahead and file it in block #13 of the flight plan. Since you've already covered yourself for ETA plus or minus 1 hour on the destination airport, the alternate must forecast at least 800 and 2 at your ETA—period. The regulation considers whether the alternate airport has a precision or a nonprecision approach procedure published, but you will be on the safe side by using the 800 and 2 rule.

In that rare (and usually not too sensible) situation when you might want to select for an alternate an airport that has no published approach procedure, you must be able to arrive over that airport at MEA (that's the lowest enroute altitude that ATC will permit) in VFR conditions and make your descent, approach, and landing in VFR weather.

When choosing your alternate, make sure that you can get from destination to alternate and still have 45-minutes' worth of fuel (a full hour is better) at normal cruise power—more on this later.

The rules to be observed when selecting an alternate airport, including the precision versus nonprecision mumbo jumbo, are shown in Table 6-2.

Now that you have jumped through all the hoops, decided that you need an alternate, and have selected an appropriate airport, for-

**Table 6-2.** Conditions and Answers to Question 2 Regarding the Selection of an Alternate Airport

| Condition | May I use this airport as an alternate? |
|---|---|
| No approach procedure published, but I can make it from MEA to the runway in VFR conditions | Yes |
| Nonprecision procedure published, forecast at ETA at least 800 feet and 2 miles | Yes |
| Precision approach procedure published, forecast at ETA at least 600 feet and 2 miles | Yes |

get all about those weather minimums; they are just for planning purposes and have nothing whatever to do with what must exist when
you get there. When you arrive at the missed approach point, only
the ability to see the runway environment and being in a position for
a normal landing govern your decision to continue to a landing or execute the missed approach procedure.

Through lack of weather observation facilities or local restrictions, some airports can't be used as alternates or require higher
ceiling and/or visibility. Jeppesen charts include this with the approach charts, and the government publications use a special symbol in the Remarks block of the approach chart to refer you to a
separate listing.

# ✛ Aircraft and Instrument Preflight

A checklist within a checklist—that's the way to make sure that once
you get into the air on an instrument flight all the gauges and gadgets
you need for aircraft control and navigation are doing their jobs, and
doing them properly. An instrument checklist religiously followed on
every flight will force you into two situations that can do you nothing
but good. First, a preflight checklist will save your pride (ever tried to
taxi away from the ramp with the airplane still tied to the ground?),
and perhaps your neck; second, a printed, orderly checklist will force
you to slow down when preparing for an IFR trip. There's nothing
wrong with hurrying, but there's a lot wrong with *being* hurried, and
the checklist will help you keep your head when all those about you
are losing theirs.

All the items on this instrument preflight are in-cockpit checks,
except two, which are of extreme importance. If you can't tell how
high or how fast you are flying, you've got real troubles, so as you
march diligently around kicking the tires and checking all the things
you normally check, pay particular attention to these two: the static
pressure source and the pitot tube. Most static ports are stainless steel
discs set flush with the aircraft skin, drilled with several small holes to
admit air pressure to the static system. If any of the holes are covered,
deformed, or plugged with paint, wax, or dirt, don't fly until you get
it fixed. The pitot system plumbing must also be free of obstructions,
including the cover you bought to keep ice and snow and house-
hunting insects out of the tube. Even with a long red streamer with
fluorescent letters warning the pilot to *Remove Before Flight,* the next

pitot cover that is taken for a short airplane ride won't be the first one! For IFR flying, you need a heated pilot tube, and the only completely reliable way to check its operation is to turn it on and feel the heat. (*Caution:* Let not thy sensing digits remain overlong in contact with the heated part, lest they come away medium rare!)

A general preflight item, which should become a habit so strong that you feel uneasy until it's done, is checking the fuel situation. Don't ever trust anybody but yourself to make absolutely sure there is as much gasoline in the tanks as you think there is and that it's the grade you ordered. Gauges have been known to lie, and line workers occasionally get distracted and forget to fill the tanks on the other side of the airplane, so the last thing you should do before climbing up to the flight deck is *check the tanks*. You may not always want them brim-full, but let your own two eyes gaze into the depths of the fuel cells and be sure you know how much is there. You'll realize a bonus from this preflight check because you'll be certain that the fuel tank caps are replaced properly and securely (the same goes for oil dipsticks). When a lid works loose in flight (or maybe was never put on properly in the first place), you can rest assured that the low-pressure area on top of the wing will aid remarkably in the rapid depletion of the tanks. There is also the problem of what-do-I-do-now? with half the fuel supply gone and no close-by alternate.

With engine(s) running and your normally used radio equipment turned on, you should begin the operational part of the instrument checklist (see Fig. 6-1).

That's a preflight planning schedule that will keep you honest in any IFR situation. Alter and adjust it to fit your airplane or the facilities you use, but remember that any one of these items could cause trouble if overlooked. To wrap it up, here's the preflight checklist in abbreviated form:

1. Check charts and equipment.
2. Check weather.
3. Select route and altitude.
4. Check distance and time enroute.
5. Check fuel required.
6. Check alternate requirements.
7. Preflight aircraft and instruments.

**Fig. 6-1**   *Instrument checklist.*

1. **SUCTION, AIR PRESSURE/VACUUM, GENERATORS OR ALTERNATORS**
Check for "in-the-green" or normal load conditions.

2. **ALTERNATE STATIC SOURCE VALVE OR SWITCH**
Check closed or normal. (If you haven't done so by now, open this valve in flight and record the difference in airspeed and altitude readings; they'll always be in error on the high—dangerous—side.)

3. **AIRSPEED INDICATOR**
Check for normal "zero" indication. (Some never go all the way to zero.)

4. **VERTICAL SPEED INDICATOR (RATE OF CLIMB)**
Check for a zero reading; if it's off, get it adjusted because you'll have to apply that correction in flight. Try tapping the case lightly to get the needle on zero.

5. **CLOCK/WATCH/TIMER**
Wind—if you still use one without a battery—and be sure that Mickey's hands are pointing to the correct numbers.

6. **ENGINE INSTRUMENTS**
Check them all in the green or at least moving that way. (Of course, you made sure you had oil pressure immediately on engine start.)

7. **RADIO EQUIPMENT (COMMUNICATIONS AND NAVIGATION)**
Turn on everything that can be checked at this point. If there's a VOR or VOT on the airport or close by, you can check the VOR receivers and possibly the DME. Tune and identify an ADF station and check for proper needle movement and relative bearing. Make sure the Loran or GPS units are setting themselves up.

8. **PUBLICATIONS, PERSONAL EQUIPMENT**
Before you leave the chocks, check once more to be sure you have all the charts and equipment you will need.

9. **ALTIMETER**
Set the altimeter to field elevation; then put in the altimeter setting you get from Ground Control or ATIS. If the hands move more than 75 feet either way, you have an altimeter that is outside the generally accepted limits of accuracy for IFR flight. If the error is less than 75 feet, don't attempt to carry the difference

forward on subsequent altimeter settings because it is too easy to correct in the wrong direction, and most of these small altimeter errors tend to cancel out over the duration of a flight. If 75 feet makes the difference between a landing and a missed approach at destination, you're pressing your luck.

10. **TURN AND BANK INDICATOR**

As you taxi, check to see that the indicator shows turns in the proper direction and that the ball moves freely in the opposite direction. If it moves in the *same* direction, call up and thank the airport manager for banking the taxiways.

11. **ATTITUDE INDICATOR (ARTIFICIAL HORIZON)**

Normal taxi speeds will not affect the bank attitude of the airplane very much, but you will notice some displacement on the attitude indicator. If the bank exceeds 5 degrees while turning on the ground, the instrument is probably unreliable for IFR flight. On those indicators that are adjustable up and down, be sure the little airplane is where it should be to indicate the pitch attitude for level flight at your normal cruise airspeed.

12. **MAGNETIC COMPASS**

While rolling straight ahead at a steady speed, check the reading against a known magnetic heading—a taxiway which parallels a runway, for example. If the deviation card shows some error, count that in your check.

13. **HEADING INDICATOR (DIRECTIONAL GYRO)**

Set this to agree with the corrected reading from the magnetic compass and make sure the card remains stationary after you make the adjustment. Recheck the heading indicator when you are lined up on the runway; it's the most accurate check you can make.

14. **DE-ICING AND/OR ANTI-ICING EQUIPMENT**

Check for proper operation (visible check of boot inflation, load-meter check for electrical devices).

15. **CARBURETOR HEAT (OR ALTERNATE AIR SOURCE FOR FUEL-INJECTED ENGINES)**

Your best friend in icing conditions—check to make sure it works as it should.

16. **AUTOPILOT/FLIGHT DIRECTOR**

Go through the operational check specified by the manufacturer, paying special attention to the autopilot disconnect and trim override features.

*(Continued)*

### 17. CLEAN-UP AND SET-UP

After checking everything, and you're satisfied that the airplane will fly, take a moment or two to clean up the cockpit. Put away all the charts except those you'll need immediately after takeoff. Before you call ready to go, set up your radios for the task at hand. If the weather is really tight, you'll do well to have one navigation receiver tuned to the primary approach aid for the departure airport; if anything goes wrong right after liftoff, you'll have one chore already taken care of and can transition to the approach situation smoothly. You should also have the approach chart readily available for the same reason. Weather not too bad? Set the navigation gear on the proper frequencies and courses for your departure.

### 18. BEFORE-TAKEOFF CHECK

Using a printed checklist or at least one of several mnemonic-type reminders, make sure all the little things are done: doors closed and locked, boost pumps on (if required), fuel selector on proper tank, etc. CIFFTRS (pronounced "sifters") is an old standard that reminds you to check:

Controls

Instruments

Fuel

Flaps

Trim

Runup

Seat belts fastened

# 7

# IFR Clearances

There is an all-encompassing rule in instrument flying, the "11th Commandment," which says, "Thou shalt not enter IFR conditions in controlled airspace without a clearance from ATC." A clearance, once accepted and acknowledged, is your authorization to proceed to a certain point, via a certain route, and at a certain altitude. With the majority of today's IFR flights under radar surveillance, the clearance becomes less a separation tool and more a means of ensuring continued safe operation when communications fail.

Clearances must conform to aviation's three-dimensional environment and therefore must always specify: (1) *a point* to which you are cleared, (2) *a route* by which you are expected to get there, and (3) *an altitude assignment.* Amendments to previously issued clearances may change any or all of these, but prior to entering IFR conditions in controlled airspace, you need all three. If the point to which you are cleared falls short of your destination airport (and this is sometimes encountered because of heavy IFR traffic), one more instruction must be added: a time to expect further clearance. The same holds true for any interruption (such as a hold) in your instrument flight.

In general, pilots are required to comply with ATC clearances (or instructions) on receipt. But the controller's use of words will provide a clue as to the timing that's expected. Suppose Center wants you to go to a lower altitude: "Barnburner 1234 Alpha, descend to and maintain 4000." You are expected to acknowledge the new clearance and then begin descending; there's clearly no urgency about this change of altitude. But when a controller says, "1234 Alpha, descend to and maintain 4000 *immediately!*" you'd better be on your way down even before you pick up the mike—that controller wants you out of your present altitude right away. When there's no traffic conflict and there's no major concern about when to descend, you'll hear, "1234

Alpha, descend to and maintain 4000 at pilot's discretion." Acknowledge the clearance and, when it suits your situation, start down. (It's wise to advise the controller when you commence the descent.)

So, when you hear:

"Descend to and maintain..."—Comply upon receipt.

"Descend to and maintain...immediately"—Do it now!

"Descend to and maintain...at pilot's discretion."—Do it when you're ready.

## ✝ Copying Clearances

Time was when a pilot had to have executive secretary shorthand skills to copy a clearance like this, delivered at machine-gun speed by a ground controller with a warped sense of humor:

> ATC clears Barnburner 1234 Alpha to the River City Airport via Victor 21 to the Maple intersection, Victor 418 Stone, Victor 25 to the Lewis intersection, direct Elmville VOR, Victor 182 Downy intersection direct, climb to and maintain nine thousand, cross the Hometown 23-mile DME fix at or above 4000, cross the Smithville 120 radial at or above 5000, cross Maple intersection at or below 6500, expect higher altitude at Lewis intersection, contact Hometown Center on 123.8 when established on Victor 21, squawk one one zero zero; read back.

There are several ways to handle a situation like this. You can develop a clearance shorthand and hope for the best, or you can submerge your pride and say, "Ready to copy; *take it slow.*" The second way is the better because it will result in fewer errors, less repetition, and more important, will contribute to a safer operation. Both parties will have a complete understanding of all parts of the clearance; mistakes of routing, altitudes, or frequencies will stand out like an oil leak on a new paint job. Before accepting an invitation to copy a clearance at a strange airport, take a few moments to listen to other clearances and study the chart of the local area—you will reduce the element of surprise when the controller reels off some obscure intersection or VOR.

Pilots must somehow acknowledge receipt of a clearance before it is considered accepted. Remember that Ground Control, Clearance Delivery, Center, Flight Service Station, or any ATC facility from which you might receive a clearance is *people,* and is subject to human error. If the controller really wanted you to proceed via Victor 47 to

Maple intersection but cleared you down Victor *21* through force of habit, the error might go unnoticed if you acknowledge the clearance with a simple "roger." Just like filing an IFR flight plan over the telephone or from the air, you can read back most of today's relatively simple ATC clearances with "numbers" alone. For example, when you're cleared "to the River City Airport via Victor 21, Victor 418 Elmville, maintain 9000, contact departure control on 123.8, squawk 4032," the readback might be simply "34 Alpha is cleared to River City, Victor 21, Victor 418 Elmville, 9000, 123.8, 4032." And the controller knows you're both in agreement on the important items.

The clearance delivery controllers at Chicago's O'Hare International (the busiest airport in the world) carried communications efficiency to a new height some years ago when they began a procedure to reduce the time required for clearance readbacks. The last item on the ATIS broadcast instructed pilots to acknowledge clearances by repeating nothing more than the assigned transponder code—for example, "Transglobal 1465, 4356." This put the monkey squarely on the pilot's back because, when a controller heard that code, it was assumed that the clearance had been received, was acceptable, and would be complied with. This procedure seems to work quite well at other airports when outbound traffic is heavy, but a controller is completely justified in requesting a complete readback. Don't argue; just read it back. When you use the transponder shortcut, be absolutely certain you have all the names and numbers straight. And just in case the controller didn't train at O'Hare, you might make your code clearer by saying, "34 Alpha copied all, 2371."

The nature of your IFR trips will have a bearing on the mechanics of copying clearances; if you frequently fly the same routes, your clearances will not deviate much from a pattern, and you'll know what to expect. But when you're trying to taxi a castering-gear taildragger in a 20-knot crosswind on solid ice, busier than a DC-3 copilot at gear-up time and wishing you had another hand and foot to help, that's not the time to say "go ahead" when Ground Control says your clearance is ready. Tell the controller to "stand by" and, when you have the airplane stopped in the runup area and your wits collected, indicate that you are "ready to copy." Now you can devote yourself 100 percent to the business of copying the clearance correctly, noting changes in routing or altitudes.

And while you're at it, why not copy the clearance in soft, erasable pencil right on the chart? It gives you a record of the clearance in the handiest possible place, available for instant reference as you fly, and eliminates another piece of paper in the cramped confines of your flight deck. You'll find plenty of blank space for this purpose on every chart.

When the controller comes to "….contact El Paso departure control on…," you know that the next item will be a communications frequency, and you can save more time by "writing" that frequency on the radio you're not using. The same technique holds good for transponder codes; when the controller speaks the numbers, "write them down" on the transponder itself. Writing on radio sets and transponders makes them the world's most expensive notepads.

Your clearance will usually come through with route and altitude just as you requested, and if you plan to fly that way again, make a note of the routing; you have probably hit on a combination that ATC will buy next time around. But if the airways you have chosen don't fit into the computer's assessment of the traffic situation, the controller will "suggest" another route. Despite what you may on occasion feel is an attempt to see how much out of your way they can make you fly, variations from requested routes are sometimes necessary. It is the controllers' expression of the best they can do for you at the time.

Here's where it's good to have the enroute chart handy to see if the suggested route is compatible with you, your airplane, and the weather. Treating the clearance as a suggestion (a strong one to be sure, but still a suggestion), *turn it down* if you don't like the looks of it. There may be a delay, but that's immensely preferable to finding yourself loaded with ice, losing altitude, and headed toward a squall line on an airway with an MEA 5000 feet higher than you can possibly fly, all because you accepted a clearance that you didn't want. *Always refuse a clearance that will fly you into trouble!* This philosophy assumes your complete knowledge of existing and forecast weather conditions along the route.

## ⌖ Cleared as Filed

On rare occasions, you will encounter a clearance as involved as the earlier example, but more than likely you will be "cleared as filed, maintain 8000, contact Hometown Departure on 123.8, squawk one zero four six." Used more and more, "cleared as filed" takes into account the destination and routing you requested in your flight plan and does away with all those confusing intersections and airways. It's a good idea to know what route you requested, since asking the controller "which way did I file?" can be a little embarrassing. And if you're not sure, you'll have to ask because the clearance is your contract with ATC for separation if the radios quit enroute. "Cleared as filed" covers *only* destination and routing; you will always be assigned an altitude. It is your choice to accept or decline, but *every* IFR clearance will stipulate the height at which you are to fly.

Anytime that you change your intended route of flight after filing a flight plan, be very suspicious of "cleared as filed." Remember that there is a delay in getting a flight plan request from FSS/Ground Control/Clearance Delivery to Center and back, and the route by which you are cleared "as filed" may be the original one. When in doubt, ask the controller for verification of the route; it may take an extra minute or two, but it might save your skin in the event of radio failure. (There it is again…when in doubt, *ask.*)

## ✝ What'd They Say?

Perhaps the best way to learn the jargon of clearances and sharpen your copying artistry is to buy an inexpensive VHF receiver and listen to the "pros" at work. (Besides, it gives you another excuse to drive out to the airport on those rainy Saturday afternoons when the alternate is to move furniture or paint the living room.) Pretend you are receiving the clearances and copy them down. Take heart as you listen and realize that even the most experienced pilots stumble over a routing now and then. When pilots claim that they've never blown a readback, don't believe anything else they say.

And it's not always the frailties of *Homo sapiens* that cause problems. Late one IFR afternoon, during the daily contest to see which controller could issue the longest clearance, an airline crew got the winner. There followed a long silence that was finally broken by the controller asking, "Trans-Amalgamated 201, did you copy?" "You'll never believe this," said the captain, "but I think we've had a pencil failure—say again all after 'ATC clears.'"

## ✝ Clearance Limits

Most of the time you will be cleared to the destination airport, but now and then a traffic conflict will not permit enough separation to clear you all the way. If it appears the conflict can be resolved in a short time, the controller may issue a clearance with a specified limit, and it might sound like this for a trip from Kansas City to Oklahoma City: "ATC clears Barnburner 1234 Alpha direct to the Butler VOR, climb to and maintain 5000, expect further clearance at 37." Although a traffic conflict has prevented controllers from clearing you all the way to Oke City, they are telling you that they're willing to accept your flight in the IFR system, that they can guarantee separation to Butler, and that further clearance will be issued before you get there. The "Expect Further Clearance" time provides an emergency backup clearance if your radios (or theirs, for that matter) fail. Should that unlikely event occur, you would be expected to hold at Butler until 37 minutes past the hour and then proceed along your route in accordance with the regulations. When there's no communications problem, controllers will issue a new clearance well before you arrive at a clearance limit.

As always, you have the option to accept or decline such a short-range clearance, but with the implication that "we'll get you to Oklahoma City as soon as possible," you're better off to accept and trust ATC to clear you along the way. Should you turn down their offer, it is likely you will wait a considerable time until the conflict is resolved. Treat a clearance limit not as an indication of ATC snubbing your route request in favor of someone else, but as a sincere effort on their part to expedite traffic flow. If the probability of solving the aerial traffic jam is slim, Center will have you wait on the ground. With a short-range clearance, at least you're in the system, on your way, and that's far better than burning up gasoline on the ramp waiting for clearance all the way.

# ✝ Other Clearances

## VFR on Top

In the beginning, the FAA created IFR and VFR. As time went on, it became clear that these two major classifications of flying conditions could be made more flexible, and "VFR on Top" evolved as one of several ways to make the airspace more useful to more people.

There are two distinct kinds of VFR on Top. There is *VFR* VFR on Top and there is *IFR* VFR on Top. You have probably used the VFR type, where you were able to stay above all the clouds, complied with visibility and cloud-clearance regulations, and were able to take off, climb, and land in VFR conditions. Although there is some calculated risk (like an engine failure on top of an overcast with nothing but the Rocky Mountains below), this is a perfectly legal situation, and for noninstrument-rated pilots, it's often the only way to get any utility out of their airplanes.

The other VFR on Top is right out of the Instrument Flight Rules and can frequently be used to advantage. It is *always* an IFR clearance, and as such, it's not something that you can do whenever you feel like it. Whether ATC actually says the words or not, VFR on Top when you're in the IFR system implies a clearance to proceed as in normal IFR conditions, but with a VFR on Top *restriction*.

Since it is a restriction, you should know the rules and conditions that apply to this type of clearance before attempting to put it to use. They are:

1. You must be *cleared* by ATC to operate "on top."
2. You must *follow the route* assigned by ATC.
3. You must *report* to ATC whenever required in accordance with IFR rules.
4. You must fly at *VFR altitudes*—eastbound, odd thousands plus 500; westbound, even thousands plus 500.
5. You are responsible (as you are at any time in VFR conditions) to see and avoid other aircraft.
6. You must be able to *maintain VFR conditions,* which means observance of all the applicable cloud-clearance and visibility regulations.
7. You should be reasonably certain you can continue your flight to destination in VFR-on-Top conditions.

So you're committed to a ball game that has two sets of rules: some IFR, some VFR. Rules 1, 2, and 3 come from the instrument book; 4, 5, and 6 are VFR-oriented; and 7 is just plain old common sense.

When is an IFR flight plan with a VFR-on-Top restriction a good deal? Anytime you have a need or desire to fly above the clouds, can maintain the altitude required to do so, and ATC is fresh out of available altitudes. But temper that need or desire with the limitations of your airplane, your passengers, and yourself; the next time a pilot gets into trouble from lack of oxygen while trying to stay VFR on Top won't be the first time.

Before you can make a rational decision about requesting VFR on Top, you should have some idea where the cloud tops are located: How high must you climb to stay 1000 feet above all the clouds? The most accurate and timely answer is found in pilot reports (PIREPs), the direct observations by the people who are or were there. (By the way, more PIREPs mean a more complete picture of the actual weather conditions, so reciprocate by making reports of cloud tops, particularly when such information is sparse or nonexistent. Early in the morning or late at night, when few other hardy souls are aloft, your reports will be of great benefit to your fellow fliers.)

The next best source of cloud heights is the Area Forecast, which will give you a general idea of the altitude at which the sky is expected to go from cloudy to clear. The information is not very specific, but it's a lot better than no information at all.

There are several situations that may prompt you to consider a VFR-on-Top clearance. Maybe there are cumulus buildups that you would like to avoid visually (ATC will allow you to deviate around them); or you might need to climb out of an icing level; or perhaps you would just rather fly high, where it's clean, clear, smooth, and cool. VFR on Top is a useful tool in your IFR kit when a higher altitude will help you and ATC has run out of available "hard" (assigned) altitudes. If you can operate within the "on-Top" rules, it is usually advantageous to seek a flight level that will reduce a headwind component or let you pick up a tailwind; that's using your knowledge to operate the airplane more efficiently.

At terminals where Special VFR is not permitted, your instrument rating can help you beat the system by obtaining a clearance to VFR on Top and then proceeding on your way, especially if a full IFR clearance is not available. This situation frequently develops when a low-topped fog bank has the terminal all but shut down. *Caution:* If you take off in weather conditions that won't permit an immediate return and landing, be sure there's a suitable airport close by to which you can go if things turn sour!

A controller who suspects that the tops are rather high and sees a traffic conflict developing above a certain altitude, may clear you "to VFR on Top, if not on top at 8000, maintain 8000 and advise." If you're still in the clouds when you get there, or don't have the required visibility, you're stuck with that altitude until the traffic conflict is resolved or the weather conditions improve. Because you now have an assigned altitude, you also have an out if the radios fail.

If the cloud tops begin to rise under you, it is your responsibility to rise with them, maintaining at least 1000 feet between you and the clouds. Higher cloud tops will force a minimum increase in flight altitude of 2000 feet to be at the proper VFR level, and it is possible that a constantly rising cloud layer will soon have you trying to operate at heights that are untenable for you, your airplane, or both. At this point, when you need all the smarts you can muster, hypoxia may be robbing you of precious judgment, and in a manner that can only be described as insidious, you don't know it's happening, and worse, you don't care! As cloud tops rise, your concern should grow; it's probably time to request a hard altitude from ATC, and nine times out of ten they will grant it. "Denver Center, Barnburner 1234 Alpha, unable to maintain VFR on Top at 12,500, requesting one zero thousand." And the reply will usually be, "Roger, 1234 Alpha, descend to and maintain one zero thousand."

But the controller cannot always assign an altitude; in a sense, you forfeit some of the ATC protection when you request VFR on Top, and the controller is obliged to fit you back into the system only if it can be done without disadvantaging other flights. If this happens (and it is rare), you must remain VFR on Top, either by climbing (seldom the best choice, especially if you are pushing your altitude limits at the time) or by holding as instructed until an altitude is open. Should you subsequently declare an emergency because of a fuel shortage or an impending performance or physiological problem, be prepared to explain why you didn't comply with the regulation that makes the pilot responsible for "all preflight information affecting the proposed flight." In other words, you should have known about the higher tops enroute before takeoff. Sorry, but that's the law, and it underscores your responsibility to be reasonably certain you can make it to destination in VFR-on-Top conditions before you request or accept such a clearance.

## Special VFR

Special VFR is usually thought of as a crutch for VFR pilots. It's the simplest of ATC clearances and can often be used to advantage by instrument-rated people as well. In contrast to the full-blown IFR clearance, Special VFR designates more a clearance "area" than a clearance limit, does not assign a "hard" altitude, and makes no provision for a specific route or for radio failure. You will be cleared into or out of the appropriate airspace, from or to a direction—north, east, southwest, etc. For example, "Barnburner 1234 Alpha is cleared out of the Fresno Class D airspace to the east, maintain Special VFR conditions at or below 3000 feet while in Class D airspace."

The heart of Special VFR is the airspace in the immediate vicinity of the airport designated as Class C, D, or E. It is there only for the protection of pilots executing IFR approaches and departures at that airport. It is controlled airspace, and you may not fly there without a clearance when conditions are less than VFR.

Because a lot of non-instrument-rated pilots are inconvenienced on days when the ceiling is adequate but the visibility won't creep past the 3-mile mark, or when they can see clear into the next county under an 800-foot cloud deck, Special VFR was developed. It is a *clearance* to proceed into or out of the appropriate airspace; the in-flight visibility minimum is 1 mile; you must stay clear of all clouds and respect the altitude limits specified in the clearance. (The altitude limitation is usually imposed to protect IFR operations at higher levels.) This is almost an IFR clearance, available to any pilot—instrument-rated or not—during daylight hours. After dark, an instrument rating is required, so why not request a full-blown clearance and enjoy *all* the protection of the system?

A full IFR clearance supersedes Special VFR; that is, no Specials will be issued when IFR traffic is inbound or outbound and a conflict is possible. This means that Special VFR folks sometimes face long delays when IFR operations are heavy. Being instrument rated puts you into the preferential group in this situation, and filing IFR will usually produce an earlier departure or arrival. At certain airports around the country, the flow of IFR traffic is so heavy and so consistent that Special VFR operations are not permitted at any time, period. Know which airports are in this group, and don't ask a controller for something that can't be done.

It's at the less busy airports that Special VFR serves all pilots most beneficially. Suppose you are waiting for your clearance at West Side Airport (tower equipped) and the minutes as well as your gasoline are wasting away because continuous IFR approaches and departures are in progress at Downtown International, 10 miles distant. If the entire area is solid IFR, settle down, cool your heels, and wait—you've no choice. But if you can depart West Side in Special VFR conditions and *very soon thereafter* fly in "regular" VFR, you may be able to get a Special VFR clearance for departure and be on your way. (*Caution:* You will likely have to refile your IFR flight plan, as ATC will probably not allow you to cancel your IFR request, take off under Special VFR conditions, and then pick up the same IFR flight plan in the air.) Conversely, you can sometimes get into a smaller airport with a Special VFR clearance when there's a long waiting list for approaches at a busier terminal nearby. If conditions are favorable, it doesn't hurt to inquire.

An inbound Special is usually not much of a problem, since you are flying toward a definite location (the airport) and various aids to navigation (radio facilities, city lights and/or landmarks, assistance from the controllers' Tower, and the like). It's the Special VFR *away from* the airport that is strewn with pitfalls for the unwary pilot; if traffic permits, you will be cleared out of the appropriate airspace, and that's as far as the controller's responsibility goes. Should the weather beyond the airport prove to be worse than expected, the controller is under no mandate to let you back in, especially if there are other Specials or IFR traffic clamoring for attention. And there you are, in IFR conditions without a clearance, can't get one, and trying to think up a good story to tell the FAA at your violation hearing!

Recognize the potential hazards of Special VFR and use it only when it can *safely* get you where you need to go. If you get in a bind and can't maintain the required conditions, don't press on, but contact ATC (Tower, Approach Control, anybody), explain the situation, and get a clearance as appropriate. Pride goeth before a fall, and when you are in an airplane, the fall usually hurteth a lot.

## Cruise Clearances

Ask any 10 instrument pilots to define a cruise clearance and you may get 10 different explanations, all the way from "maintain cruise airspeed" to "descend immediately to minimum altitude." A very useful tool of IFR operations, the cruise clearance carries a number of implications which bear on the efficiency and safety of getting to your destination, and it must be thoroughly understood to be used properly. When an ATC controller says you are "cleared to cruise," it means:

1. You are cleared *direct* to the specified navaid or airport.

2. You are cleared to fly at any altitude from that specified in the clearance down to the appropriate IFR minimum altitude. A word of caution here: "Appropriate IFR minimum altitude" could be MEA if you're on an airway or the published MSA when you *know* you're within 25 miles of the appropriate approach aid. When in doubt, stay at the altitude specified in the cruise clearance and ask the controller when it's safe to descend. Under certain conditions, the controller can provide a minimum vectoring altitude. Don't guess: Stay high, and *inquire.*

3. You may leave the cruise altitude at your discretion for a lower altitude, because your altitude clearance is for a vertical slice of airspace, described in item #2. If you *do* report leaving an altitude, you may not climb back up to that level without a fresh clearance to do so.

4. You are the only *IFR* flight at cruise altitude and below between you and the navaid.

5. You are cleared for the published approach of your choice on arrival at the airport.

6. You are responsible to advise ATC (the facility will be specified in the clearance) when landing is assured or upon missed approach, depending on the situation.

Cruise clearances were designed primarily to accommodate the final segment of an IFR flight and are generally used at locations that have no tower or approach control facility. (A cruise clearance issued before takeoff differs from a normal one only in the definition of altitude; see item #2.) A very flexible instrument of traffic control, this clearance allows you to begin descent at your discretion, choose an altitude best for the situation (turbulence, headwinds, icing, inoperative cabin heaters, or other fun things), and ideally break out of the clouds sooner than you might otherwise. When you get to the appropriate minimum altitude and find yourself in VFR conditions, there are several ways you can complete the approach.

First, if you want the practice, go ahead with the full IFR procedure (it will take a little longer, but there's something to be said for the IFR separation this offers, plus the fact that it will keep you from getting lost or landing at the wrong airport!). The second method, if *good* VFR exists in the terminal area, is to cancel your IFR flight plan, removing you from the system (although most radar controllers will provide traffic advisories as long as their workload permits). The third and fourth possibilities are the Contact and the Visual approaches, treated in detail in Chap. 15, Instrument Approaches.

## "Cleared for the Approach"

Regardless of the areas in which you fly IFR, there comes a point in every trip when the controller turns you loose and expects you to complete the published procedure entirely on your own. The phrase most commonly used for this purpose is "cleared for the so-and-so approach," and it is your cue to fly the procedure as published (certain methods of shortcutting the full procedure are outlined in Chap. 15, Instrument Approaches). But in more and more real-world situations, that phrase won't be heard until you have been assigned a radar vector intended to intercept the final approach course—a definite timesaver and one of the very good things about radar control.

But be aware of an altitude trap and the rule that was written to keep you from falling into it: Whenever you are operating on an unpublished route (as with RNAV or when flying direct from a VOR to intercept a final approach course) or flying in accordance with a radar vector, you must maintain the last assigned altitude until a lower one is issued by ATC or until you are on a segment of a published route or approach procedure which has a published minimum altitude. In other words, don't descend until you know (from the numbers on the chart) that it's *safe* to do so.

# ✝ The Last Word in Clearances

There are, we hope, no controllers who issue capricious clearances and no pilots who willfully and knowingly disregard or deviate from the provisions of an ATC clearance they've received. But controllers and pilots are humans, and as such, they are subject to misunderstandings relative to what the other person really said. As the result of several major accidents involving misunderstood clearances, the rules have been expanded to include the following proviso under the title "Compliance with ATC Clearances and Instructions": "If a pilot is uncertain of the meaning of an ATC clearance, HE SHALL IMMEDIATELY REQUEST CLARIFICATION FROM ATC." When in doubt, ask.

A South Carolina charter pilot, who talked just about as fast as he flew his old Twin Beech (that means *s-l-o-w-l-y*), found himself approaching the New York terminal area one day at the height of the morning rush hour. The headphones were alive with a steady stream of clearances and instructions, some of which were certainly directed toward the Twin Beech. Every time he'd try to respond, or ask about a communication, the South Carolinian would be firmly "stepped on" by some fast-talking Yankee.

Following a clearance that appeared to have some significance for his flight, our pilot waited for a break and then drawled, as only a South Carolinian can drawl, "Laguahdia Approach, this heah's Beechcraft seven foah Victah...was what y'all just said to me anythin' impawtant?"

Seven foah Victah got preferential treatment from that point on!

# 8

# Communicating in the IFR System

There are a number of ways you can make your IFR communications more efficient, one of which is to get rid of the habit of prefacing every transmission with "ah." There is probably some psychological theory underlying the several types of ahs, which range from the student pilot who truly doesn't know what to say and uses ah to fill a verbal void to the 20,000-hour airline captain whose well-practiced ah is pitched at least two octaves below anyone else's and is a hallmark of accomplishment. There's nothing illegal, immoral, or fattening about using your own ah, but it takes time that can be better spent in meaningful communication. If you are a habitual aher, don't do it when you're IFR in the New York area. By the time you've finished ahing, New York Approach Control will have cleared three airline flights for approaches to JFK, handed off two Bonanzas and an Apache to LaGuardia Tower, and coordinated a Civil Air Patrol search for a sailboat missing on Long Island Sound! As terminal areas get busier and until data link systems are in common use, it is incumbent on every pilot—especially those using the IFR system—to become better managers of their communications.

Airborne communications problems have not escaped the attention of the wonderful world of research. Someone conducted a study a while back with the aid of a number of in-cockpit recorders, taping transmissions to and from the aircraft as well as pilot responses on the flight deck. The researchers found that in almost every case radioed instructions from the ground were followed by one of these phrases between pilots (listed in order of usage):

1. "What'd he say?"
2. "Was that for us?"
3. "Oh, shucks."

Practice communication management right from the start. When you turn the radio switch to the ON position, know ahead of time where the volume knob should be for normal reception. If there isn't some kind of mark on the knob, make one. On those radios provided with a manual squelch control, adjust it to a point just short of where the noise begins, and you've set up your transceiver properly for the first transmission. You can usually listen to others talking (or the ATIS broadcast) to help adjust your receiver before you talk to anyone.

# ✝ Tips for Talking

Here are several things you should do (and some things you should *not* do) to improve the quality of your communications in the air:

1. Always listen to be sure the frequency is clear before starting a transmission. You will often hear just one side of a radio conversation, so take that into consideration. If a controller asks somebody else a question, but "somebody else" is too far away for you to hear the answer, allow a reasonable length of time before you begin. You'll probably hear the controller's acknowledgment of the other pilot's answer; use that as your go-ahead signal.

2. Before you transmit to anyone at any time, know what you want to say before you press the mike button. It's not necessary to abbreviate your words as they do in the movies, but do compress the message so that you get your point across with the minimum number of words.

3. Don't make transmissions that are unnecessary. This should not preclude a friendly chat with the controllers when they aren't busy and exhibit a willingness to pass the time of day.

4. Don't click the mike button to acknowledge a transmission. To a controller, all clicks sound alike, and the instructions will have to be repeated for confirmation. If you're too busy to acknowledge (and this is not an isolated circumstance, especially IFR), it's better to comply now and acknowledge a few seconds later when you have the time.

5. If you intend to fly a lot of single-pilot IFR in busy areas, invest in a headset/boom mike and a yoke-mounted mike switch.

6. Hold the mike close to your lips and speak in a lower-than-normal tone and volume (if the person seated next to you in the airplane can hear you talking, you're probably talking too loudly). This will help eliminate engine and aircraft noise, making your transmission much more readable. Most aircraft microphones "clip" the peaks of volume and pitch, and shouting just makes a bad situation worse.

7. After you have identified a VOR, turn off the audio side of the VOR receiver. The noise and occasional conversation on the VOR frequency will serve only to distract you from more important business. Tune, identify, and turn off the sound.

8. Never, never sacrifice the control of your aircraft for the sake of talking on the radio. If you didn't know better, you'd sometimes think that Tower controllers were watching your takeoff with binoculars so they can ask you for your first estimate when they see you reach for the gear handle! The same philosophy applies to the required report on executing a missed approach—get the airplane on its way, and when everything is completely under control, let somebody know of your intentions. *Fly the airplane first!*

# ✝ Be a Conformist

You must realize that when you file and fly IFR, you become part of a highly structured and organized system intended to provide the fastest, safest service to all comers. You cannot be denied access if you are qualified, so your task is to fit into the system as smoothly as possible, going along with the established procedures, becoming a round peg when ATC wants to fit you into a round hole (assuming that it's a *safe* round hole!). Communication represents a key element of the system; no matter how distinctive you like to be on the air, you must become a "system communicator." There are two parts to the problem: First, knowing exactly *what* to say (and no more) and, second, knowing *when* to say it.

About this matter of your aircraft identification—except for those worshipers at the altar of distinction who obtain nonstandard registration numbers, nearly all civil aircraft in the United States consist of four numbers and a letter suffix. Remember that controllers will usually be mentally juggling several aircraft call signs at once, so on your initial transmission, use just the "last three" of yours to get their attention. Suppose your full set of numbers is 1234A; say "34 Alpha" the first time and provide the make and the rest of the call sign after you've established communications. If the controller answers with all of your numbers, you should use the full identification in subsequent transmissions because there may be another aircraft on the frequency with a similar call sign. If there are *two* Barnburners with numbers ending in 34A being worked by the same controller, confusion can reign supreme if both parties insist on acknowledging with just "34 Alpha." Use of the full identification completely eliminates this problem. However, if yours is the only one that comes close to 34 Alpha, the controller will probably shorten the call sign; then you may respond with the abbreviated number.

Clearance amendments that involve rerouting and other major changes should obviously be read back to ensure complete understanding by both pilot and controller. More specifically, any clearance or instruction that involves a new heading or altitude requires a readback. "34 Alpha, turn right heading one five zero, descend to and maintain 6000" should elicit this response from the pilot, "34 Alpha, right one five zero, cleared 6000, leaving 7000."

In that last exchange, notice the controller's care to place "and maintain" between "to" and "6000." There's good reason for this, since "to" can be easily mistaken for "two," and the possibility of the Barnburner pilot interpreting the new altitude as "two six thousand (26,000)" becomes a potential source of confusion. When controllers take pains to make *their* half of such a communication clear and unmistakable, it behooves pilots to respond in like manner. By reading back the new clearance as in the preceding paragraph, the controller heard exactly what was expected—the phrases and numbers that had just been transmitted and in the same order. That will surely be more understandable.

In addition to fostering more efficient communication, this technique (that is, reading clearances and instructions in the same order as they are transmitted) is based on regulation and good practice: Pilots are required to acknowledge a new clearance ("cleared 6000") and to report vacating an assigned altitude ("leaving 7000"). So, when you are established in level flight and then cleared to climb or descend, develop the habit of responding the same way every time, as in: "34 Alpha is cleared 9000, leaving 4000." You've acknowledged and reported in one fell swoop.

As you cruise about the airspace, you'll hear many variations on the theme of altitude reporting, such as "two point three" and "five point oh," but there's no altitude message that comes through as loud and clear as "two thousand three hundred" or "five thousand." Say it right the *first* time and save the time it takes to "say again."

Another opportunity to streamline IFR communications arises when you are handed off to another controller during a climb or descent. The initial report should include altitude information, but remember that the acquiring controller has been advised of your climb or descent and is most interested in confirming where you intend to level, not where you are now. Good practice suggests that you say merely, "Minneapolis, 34 Alpha climbing [or descending] 8000." 'Nuff said. A controller who needs to know your present altitude will ask.

If you receive a new altitude clearance from the *same* controller while you're climbing or descending, reply "34 Alpha now cleared 4000." No more words are required to confirm the clearance.

Just as excess verbiage seems to clog communications channels, pilots and controllers who are not positive and assertive often cause problems. Questions seem to beget questions:

PILOT: Fort Worth, this is Barnburner 34 Alpha, did you want me to turn right to a heading of two five zero to intercept the localizer or was it a left turn?

CENTER: 34 Alpha, Fort Worth, did you copy right to two five zero?...ah, negative, that's a left turn, I say again, a left turn heading two five zero to intercept. Did you copy?

And sometimes, more questions follow. How much better, when you're unsure, to be *assertive:* "Fort Worth, Barnburner 34 Alpha, *say again.*" Elimination of confusing questions is a good enough reason for having "say again" in the aeronautical vocabulary.

The use of radar transponders introduces an opportunity to save time, too. When a controller requests that you squawk "ident," you shouldn't even think about using the microphone; just press the "ident" button and let the black box do the talking. There's no need to tell the controller that you have responded because your target on the scope will "bloom," and when you're advised "in radar contact," all you need reply is, "Roger, 34 Alpha." The same is true of a code change. For example, if a controller directs you to squawk one three six four, your communication should be just that—turn the knobs until the proper code appears and wait for an acknowledgment. In those rare cases where the change isn't noticed right away, you'll be asked to confirm that you have made the change. The important thing is that you didn't clutter the frequency with an unnecessary exchange at the outset.

In passing, it's interesting to note that the term *squawk* is an outgrowth of World War II terminology that labeled the brand-new military transponders "parrots"; they replied to a coded electronic message just as the raspy-voiced green birds do. On a day when everything has gone wrong and you need an outlet for your emotions, a Center request to "squawk ident" provides a golden opportunity. Pick up the mike and, in the screechiest voice you can generate, make like a parrot: "Ident! Ident! Ident!" (You should expect some sort of nasty retaliation from ATC, but this little exercise is guaranteed to relieve your tensions.)

# ✝ Say Only What's Needed

There is a standard procedure for communicating when you are handed off from one Center sector to another or from one Center to the next. You must realize that, before controllers request that you "contact Cleveland Center now on 124.7," they have contacted the acquiring facility by land line (all Centers and their sectors are linked by phone lines), confirmed that you are in radar contact, and asked what frequency you should use. The subsequent controllers therefore know who you are, where you're going, and really only need confirmation of your altitude, the most important ingredient of safe separation at this point. So, when you check in with the new facility, you should say merely, "Cleveland Center, Barnburner 1234 Alpha, 8000." If they want to make absolutely sure, they may require you to squawk "ident," but they are not obligated to advise you "radar contact." You may assume you've been radar identified unless you're advised otherwise—another case of the system eliminating needless conversation. If the handoff occurs while you are climbing or descending, make this information part of your report: "Cleveland Center, Barnburner 1234 Alpha, climbing [or descending] 8000."

If it hasn't happened to you yet, it will: "Hometown Unicom, this is 1234 Alpha; I'll be on the ground in 10 minutes, will you call my wife and tell her to pick me up at the airport?" Followed by, "Roger, 1234 Alpha, we'll be happy to make that call for you, but it will be long distance; this is Albuquerque Center." By switching back and forth from one radio to the other, you have transmitted on the wrong frequency! Using both radios can cause a great deal of confusion; eliminate it by using only one of your transceivers; the "other" radio should be considered a standby unit. In addition to preventing you from talking to the wrong people, this procedure takes one more monkey off your back, that of figuring out which radio is the right one to use. When you're in the IFR system, anything you can do to decrease your mental workload is good.

# ✝ Play the "Numbers" Game

One of the most disturbing and time-consuming communications situations takes place on the ramp at any busy tower-controlled airport, where the ground controller reels off the active runway, taxi instructions, and the altimeter setting to a departing aircraft and then hears, "Barnburner 1234 Alpha, ready to taxi." Faced with this set of circumstances, the controller has little choice but to go through the whole bit again, and the needless repetitions build into frustrating numbers in the course of a day. When you get the engine started, turn on the radio, monitor Ground Control, and listen. Unless you're the only pilot getting ready to go, you can copy the appropriate instructions, and when you are ready to move out, save tempers and time by saying, "Barnburner 1234 Alpha, ready to taxi *with the numbers*." (If you're on a large airport with several locations from which aircraft might taxi for takeoff, don't make the controller guess where you are. State your position in the original call: "Barnburner 1234 Alpha at Acme Aviation, ready to taxi with the numbers.")

When approaching for landing, the same principle applies; you can usually monitor the tower frequency from a considerable distance, learn what runway and traffic pattern are in use, and tell the tower you "have the numbers" on the initial call: "Downtown Tower, Barnburner 1234 Alpha 6 miles northwest with the numbers, will call you on downwind for Runway 36." If the controller would prefer some other pattern entry or another runway, you'll be advised. The beauty of this procedure is that the controller will usually only need to acknowledge your thoughtful, preplanned transmission with "roger, 1234 Alpha," again making the most of communication time.

Both of these situations—before taxi and prior to landing—have been vastly improved wherever Automatic Terminal Information Service (ATIS) is installed. Take a few seconds to took up the ATIS frequency and get all the pertinent airport information before reporting "ready to taxi" or before contacting the approach controller. There's just no excuse for controllers having to spend time reciting the active runway, altimeter setting, and other pertinent information when it is waiting for you on a continuous ATIS recording. It's even more important for IFR operations, since the ATIS broadcast usually includes weather, winds, and the approach procedures in use. At most large terminals, ground controllers don't have the time to cater to uninformed pilots and will refer you back to ATIS if you call for taxi instructions without the current information. "Kennedy Ground,

Barnburner 1234 Alpha at Gate 4B ready to taxi" would no doubt be rejoined crisply with, "34 Alpha, information Bravo is current," and the controller will go on about more pressing business, leaving you to find out for yourself what information Bravo is all about.

Inbound to a busy airport in solid IFR, there's still time to tune the ATIS frequency and get yourself set for the approach segment of your flight. Granted, doing this requires listening with each ear tuned to a different frequency, and ATC instructions certainly take precedence. But if you start listening to ATIS far enough out, you can pick up portions of the broadcast in between the calls from Center (use the audio switch to cut out ATIS when you hear your call sign on the primary radio), and by the time you are handed off to Approach Control, you should have the complete message.

When you're trying to obtain ATIS information as much in advance as possible, altitude always helps (another good reason for flying high whenever you can). Here's another technique you can use to increase reception distance: Open the squelch control, put up with the static for a couple of minutes, and you can often sort out the important parts of an ATIS broadcast much farther away from the terminal than you thought was possible. (The TEST position on a radio with automatic squelch control will produce the same result.)

When the terminal weather is very good, so good that it's insignificant to instrument pilots, there will likely be no mention of it on the ATIS broadcast, or perhaps the simple statement, "Weather is VFR." But when the meteorological situation is changing so rapidly that the controllers can't keep up with it, the ATIS mill sometimes contain the words "weather will be issued by Approach Control," and you should expect to get the good (or bad) news from that facility at the appropriate time.

Good pilot practice calls for you to inform Approach Control or Tower on initial contact that you have received "information Foxtrot" (or whatever is current). To illustrate, when Center hands you off to Approach Control, the exchange should sound something like this:

CENTER: 1234 Alpha, descend to and maintain 5000, contact Bay Approach on one two five point three.

YOU: Roger, 1234 Alpha cleared 5000, leaving 8000, Approach on one two five point three.

YOU AGAIN: Bay Approach, 1234 Alpha descending 5000, with Foxtrot.

And the Bay Approach controller will figure that you're a pilot who knows which end is up.

# ✠ Getting Around on the Ground

Proper communications discipline becomes more important as an airport becomes busier and reaches its peak at those terminals equipped with a Clearance Delivery facility. In addition to listening to the ATIS broadcast, you'll probably be required to have your IFR clearance before taxi. Besides the obvious rebuff you're going to get if you contact Ground Control without a clearance, consider the unruffled ease with which you can copy when you are sitting calmly on the ramp, nothing competing for your attention except getting the clearance right the first time. If you have a good strong battery (don't try this on a cold day when the brass monkeys are heading south) and notice a long lineup at the departure end of the runway, you might consider switching on and getting your clearance before you even turn a prop blade. In the face of a long delay, this can save a lot of engine ground-run time. When you have filed for a quick turnaround, it's possible to get even farther ahead of the system by contacting Clearance Delivery as you taxi in after landing.

Whether it's Clearance Delivery or Ground Control, you can help them dig your flight plan out of the computer by providing some basic information on the first call: "Washington Clearance, Barnburner 1234 Alpha, IFR to Saginaw." Knowing your destination and that you're an IFR flight gives the controller something definite to look for.

Ground controllers will set you up by asking, "Barnburner 1234 Alpha, I have your clearance, ready to copy?" but knowing that you are monitoring the frequency, some Clearance Deliverers will just let 'er rip, ready or not! The radio suddenly comes alive with, "Barnburner 1234 Alpha is cleared as filed to the Saginaw Airport, maintain one zero thousand, etc., etc., etc." Remember that Clearance Delivery is there to relieve congestion, and you shouldn't be on the frequency unless you're ready to copy, so *be* ready; have your charts spread out to visualize route changes and all the other good practices germane to clearance copying (see Chap. 7, IFR Clearances).

Now that you have your clearance and are ready to drive the Barnburner to the other end of the aerodrome for takeoff, you may face a navigation problem that is more complicated than any airborne situation. As airports have grown in size and complexity, the number of taxiways, outerbelts, innerbelts, crossovers, and switchbacks has increased to the point where you are just about ready to concede that there is no way to get to the runway from here! An old adage of the flying business applies here: When in doubt, swallow your pride and ask. Sure, the airline captain who flies into a particular airport six times a week knows the taxi routes inside out, but ground controllers are quite aware that "first-timers" are going to experience difficulty when they are cleared "to Runway 14 via Charlie and Mike, hold short of Juliet before crossing to the inner parallel, give way to the Beech 99 approaching from your right as you cross November." And Ground is more than willing to help you find your way, but you do need to ask. On a completely unfamiliar airport, your first task is to break out the approach plate (or a separate taxi chart for the superlarge terminals; see Fig. 8-1) and, from the airport plan view, figure the most likely taxi route; trace it as Ground Control reads it off, and you should be able to make it on your own.

But suppose you can't find the chart or the taxi clearance is so confusing that you don't even know which way to turn as you come off the ramp. Here's where you admit to the controller that you are unfamiliar with the airport and politely ask for directions. The answer may be humbling because, nine times out of ten, there just happens to be an airliner conveniently located so that you will hear (in a condescending voice), "Roger, 34 Alpha, turn right on the taxiway straight ahead of you and follow the United 747 to the active runway." So, tuck in under the tail of the airliner, and it's "whither thou goest" from here to the end of the runway. Don't be chagrined because the copilot is sitting up there with an airport diagram, telling the captain which way to turn to make sure *they* are heading in the right direction.

Some air terminals are notorious for the curt manner with which their controllers operate, and others are equally famous for the complete and willing cooperation that comes from Ground Control. In either case, you certainly have a right to ask, and 'tis a far better thing to find out from them which way to turn than to taxi onto an active runway and find yourself staring down the intakes of a just-landed MD-88.

Once you are cleared by Ground Control to the active runway, you are obligated to remain on that frequency until you arrive at the hold-short line. If you want to depart momentarily to listen to ATIS or call Unicom, be sure to check off with Ground Control and report back. Upon arrival at the end of the runway or the end of the line. it's prudent to monitor the tower frequency for takeoff clearances and restrictions being issued to other similar type aircraft. Just another way to stay one step ahead by getting ready for the next segment of your flight before you actually get there.

When yours is the only airplane waiting to leap into the blue (or the gray, as the case may be), the switch to Tower frequency should

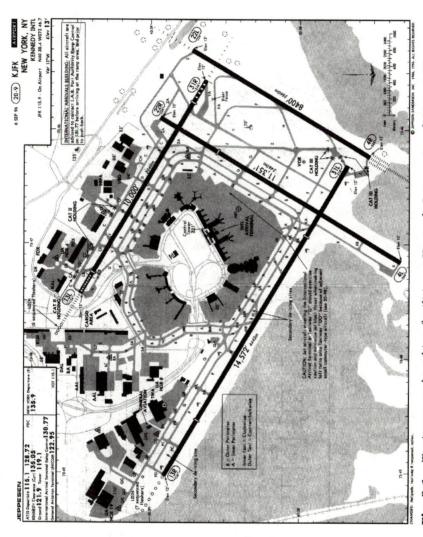

**Fig. 8-1** *Taxiways and parking facilities at Kennedy International Airport, New York. (Copyright Jeppesen Sanderson Inc., 1986, 1996, Not to be used for navigation. All rights reserved.)*

be made just as soon as you have all the knobs, switches, levers, and handles in the proper positions for takeoff, and all you need say is, "Possum Kingdom Tower, Barnburner 1234 Alpha, ready on one four." Especially at busy airports, including the runway number helps the controller sort out the departure requests. When a takeoff release is obtained from Approach Control or Center, Tower will advise, "Maintain runway heading [or other appropriate instructions], cleared for takeoff," and away you go.

Tower controllers must always coordinate IFR departures with either Departure Control or Center, which means a communications lag, however slight. You can often circumvent this delay (assuming no lineup of waiting aircraft, in which case you must wait your turn) by advising Ground Control as you leave the ramp that you will be ready for takeoff when you reach the runway. This gives Tower a much more positive departure time to pass on to the other ATC facilities, and your takeoff clearance will probably be waiting for you when you switch to Tower. If you do this, don't delay getting to the runway because a phone call will have been made informing someone else up the line that 1234 Alpha can be off the ground in 2 minutes. (This procedure also means that you must figure out some way to accomplish your pretakeoff checks before you reach the head of the runway. If you've any doubt that you can do this thoroughly and safely, taxi to the run-up area and don't call "ready to go" until you finish the checklist.)

## ✝ Position Reports: A Lost Art

One of the most pleasant sounds to filter through your speaker on an IFR flight is the phrase "in radar contact." It will usually happen right after takeoff, when the departure controller sees you on the scope, and in most parts of the country, you will continue in radar contact all the way to destination. It's comforting to know that someone down there knows who you are, where you are, and where other traffic is; but there's an additional feature that has done more to clean up the communications clutter than anything else. When you are advised that your flight is "in radar contact," controllers don't want to hear from you, since they know your position, groundspeed, and ETAs more accurately than you do. They don't want you to speak unless spoken to; as far as ATC is concerned, IFR pilots—like little children—should be seen (on radar) and not heard (on the radio).

Full-blown position reports are definitely taboo when in radar contact, but there are several other circumstances in which ATC would like to hear from you, mostly with confirmation of your actions which bear on safe separation from other IFR flights. Therefore, you should report without request when these situations exist (talk to the ATC facility with which you are in contact at the time, or as instructed):

1. Report the time and altitude reaching an assigned holding fix or clearance limit.

2. Report when leaving an assigned altitude. (Notice that a report is not required or even desired when reaching the new altitude. If ATC wants to know when you get there, a separate report will be requested.)

3. When you leave an assigned holding fix or clearance limit, let ATC know what time you departed.

4. On an instrument approach, you should report when you pass the Final Approach Fix inbound to the airport. ATC will always let you know to whom you should report.

5. You must report when executing a missed approach—again, you'll be told whom to call.

6. If you have given ATC an estimate for a subsequent reporting point, and it appears that your estimate will be in error by more than 3 minutes either way, let the controller know.

7. When you're operating IFR with a VFR on Top restriction, you may fly at any altitude you choose (VFR hemispheric altitude rules apply), but advise ATC if you make a change.

Some communications situations demand interpretation on the part of the pilot; ATC people use rather severely abbreviated phrases when time is at a premium. Such was the case one extremely busy morning when a Center controller needed confirmation of the height of a particular flight and transmitted, "Trans Global two thirty four, say altitude." The reply came back, "Altitude." In a slightly more stentorian tone, Center tried again: "Trans Global two thirty four, say *altitude!*" The pilot was still feeling frisky and replied again, "Altitude." Taking the measure of his airborne adversary and displaying more than the usual amount of savoir-faire, the controller broadcast calmly, "Roger, Trans Global two thirty four, say 'canceling IFR.'" Only a few seconds separated that transmission from, "Center, Trans Global two thirty four is level at one zero thousand."

On occasion, and particularly at lower altitudes over mountainous terrain, you will be asked to contact another ATC facility on a new frequency, and when you dial in the proper number, nobody answers. When this happens, wait a minute and try again. If the third attempt elicits no response, return to the previously assigned frequency and advise the controller of your plight. Nine times out of ten, the next controller heard you calling but was too busy with other traffic to answer; or perhaps you are just a bit beyond the range of the controller's transmitter.

Whatever the reason, don't go through a lengthy dissertation on what has happened; just say to the previous controller, "Memphis Center, Barnburner 1234 Alpha, *unable* Kansas City on one two three point eight." The magic word is "unable," and it identifies the difficulty right away. After advising you to "stand by," the controller will call Kansas City on the landline and ask if they heard you. Sometimes the new frequency is a bad one, and you'll be assigned another, but more than likely, Memphis will ask you to try again, maybe 5 minutes from now, when you'll be close enough to hear Kansas City.

Don't get uptight and call over and over; continue on your way, and after a reasonable time, try to establish contact. Even if your radios have quit, there are ironclad rules to follow (see Chap. 12, Communications Failure); of course, before you launch into those procedures, try to contact a Flight Service Station, a Tower, or any handy ATC facility.

# ✝ Letters and Numbers

The International Civil Aviation Organization (ICAO) phonetic alphabet is the rule book when it comes to proper pronunciation of both letters and numbers. Some of the letters may seem a bit unwieldy to you, but they've been chosen for their universality; theoretically, pilots from any country in the world can make themselves understood using these phonetic words. But there are problems, even in the US of A. For example, you will seldom hear pilots from the Midwest say *oss-cah,* as Oscar is spelled phonetically in the book; those pilots are bound to drawl *oss-curr* from now 'til the end of time. Their downeast counterparts will likewise pronounce Sierra *see-air-er,* and the sodbusters who call 'em like they see 'em will always use *ly-mah* as in beans instead of *lee-mah* as in Peru.

Numbers come in for their share of abuse, too, the most easily muddled combination being "five" and "nine." To eliminate confusion entirely, the communications expert would have us use "niner," which does the job very effectively. (A man who suspected his wife of hanky-panky with an airline pilot had his suspicions confirmed when, confronting her with the evidence, he demanded to know if indeed she had been indiscreet. "I've told you niner thousand times, *negative!*" she replied.)

Transiting the IFR airspace can be a trying experience for those who are not familiar with "what's happening," but we all had to start somewhere. If you are not a smooth communicator, who is able to become part of the system with a minimum of radio conversation, buy a VHF receiver and listen; learn the jargon of instrument flying, anticipate the instructions coming up, and play the game of thinking ahead so that you can impart the most information with the fewest words. You'll endear yourself to ATC, save time for all the other pilots who have important things to say, and brand yourself a "pro" in the process.

# 9

# VOR Navigation

Keeping the Course Deviation Indicator (CDI) centered enroute is a fairly simple task, and by always placing the airway course under the index, you can't go too far wrong. But when it's time for a VOR approach or a holding pattern, you've got to know what the receiver is telling you and how to orient yourself to the desired radial.

All VOR displays have this much in common: Each one is composed of: (1) a CDI or, in everyday language, a left-right needle; (2) a TO-FROM indicator, which solves the question of ambiguity; (3) an Omni Bearing Selector (OBS), which sets the stage for interpreting the information from the other two. When the OBS is set on a particular course, the CDI will *always* tell you whether you are on, to the left of, or to the right of the radial that the OBS setting represents. The TO-FROM indicator furnishes additional information about whether that course (the one on which the OBS is set) will take you TO or FROM the transmitter.

Because of this orientation to a specific course, it is imperative that you begin interpretation of what you see on the panel by getting the airplane lined up on a heading the same as, or at least close to, the course you have selected. You can do this by physically turning the plane to that heading or, more likely, by imagining yourself flying in that direction. When this basic condition is met, the VOR receiver indications will always tell the truth. If you fly a heading that agrees with the course set into the OBS, you will proceed toward the station if the indicator shows TO or away from the station if it shows FROM. If the CDI is centered when heading and OBS are close together, you can be sure that you are ON that particular radial; if it's displaced to the left, you must fly left to get to the radial, or fly right if the needle lies on the other side. Unless you are turning to intercept a radial or just entering a holding pattern, the only difference between the OBS setting and your heading should be drift correction, and it will seldom be a very large disparity. (Technically, the "truth" on the VOR indicator exists on any heading up to 90 degrees either side of the selected course, but you'll do yourself a navigational favor by always imagining the airplane's heading in agreement with the OBS. If it's necessary, turn the airplane until the numbers agree and proceed from there; it's better than getting completely disoriented.)

## ☩ Radials, Radials, Radials

If you don't speak, read, and write the language of radials, there's no time like right now to fix it firmly in your IFR thinking because the entire VOR navigational system is based on courses *away from the station*. There is only *one* line on the chart for each numbered radial associated with a particular VOR station; whether you are flying it outbound or inbound, holding on it or crossing it, a radial is always in the same place. The only possible complication lies in the reciprocity of the numbers: Whenever you are proceeding *outbound,* your magnetic course (and heading when there's no wind) will be the same number as the radial; turn around and fly *inbound* and you must mentally reverse the numbers and physically reverse the OBS setting so that your course is now the reciprocal of the radial. Be that as it may, you are still flying on a *radial*; it hasn't moved or changed at all.

## ☩ Putting VOR to Work

There are four basic problems you will encounter in everyday use of VOR. Whether you're holding, executing an approach, or navigating enroute, a thorough knowledge of these four will enable you to handle any situation. They are:

1. Determining what course will take you direct to a VOR.
2. Determining your position in relation to a specific airway or radial.
3. Identifying an intersection (or crossing a particular radial).
4. Determining a wind correction angle that will keep you on course.

The first case is frequently put to use when you are cleared "from your present position *direct"* to a VOR. The procedure is simple; after tuning and identifying the station (don't *ever* forget this step!), rotate the OBS until the left-right needle is centered and the ambiguity indicator shows TO. The number that now appears under the OBS index is your course to the station (inbound, it will always be the *reciprocal* of the radial you are on); turn to and fly that heading, applying drift correction as necessary to keep the needle centered, and that's all there is to it.

The second case starts off the same way; when you are requested to "intercept and fly outbound on the 320 radial," make certain you have the right station tuned and select 320 on the OBS (since you are outbound, desired course and radial are the same). If you're not on that heading or close to it, turn to 320 degrees to satisfy the first condition of VOR orientation. Assuming that when you received the clearance you were already northwest of the VOR, the TO-FROM indicator should settle on FROM. (If you're located somewhere else, keep flying on a 320 heading; you'll get a FROM sooner or later.) So far, so good; now to the heart of the problem: Where are you in relation to the 320 radial? If the CDI centers, you're there, and because you turned the airplane to bring the heading into agreement with the OBS setting, a deflection of the needle to either side will tell you which way to turn to get on course. Needle left? Fly left, and vice versa. The same procedure applies when you're cleared to fly a radial inbound (don't forget to reverse the numbers), except the TO-FROM indicator will read TO.

The third situation comes up frequently, such as when you're checking groundspeed, temporarily out of radar contact, timing a holding pattern, or asked to report at an intersection. Reporting points are usually the well-defined crossing of two or more airways or radials. Whether you're asked to report at a named intersection or when crossing a certain radial, there's a technique that will work every time. For example, if you are proceeding northwest from the Peoria VOR on Victor 434 (Fig. 9-1) and the controller requests a report passing the VICKS intersection, you have a choice of either the 061 radial of the Burlington VOR or the Bradford 247 radial as your cross-reference. The Bradford radial seems best because you will be closer to that station and the bearing will be close to a wingtip position. (Radials indicated by arrows are recommended, but you can use any combination you like, as long as your position is not beyond the VOR changeover point and the angle is not less than 30 degrees.)

When you tune and identify Bradford, place the desired radial (247) in the OBS and, when the CDI centers, you have reached VICKS intersection (maintaining the centerline of Victor 434 all the while, of course). Until you get to VICKS, the needle will remain to the right of center, so here's the rule: When the "side" radial is set in the OBS, you have not arrived at the intersection as long as the CDI is deflected

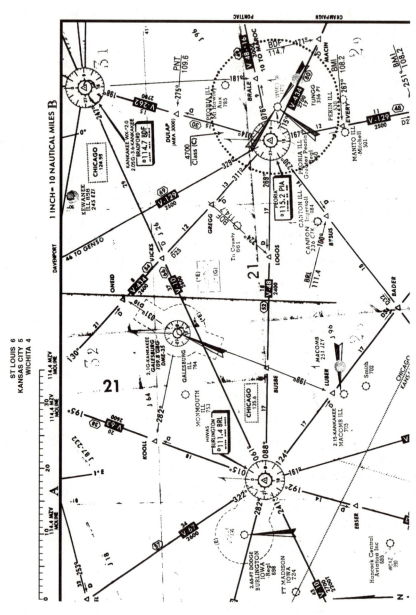

**Fig. 9-1** *Navigational facilities forming the VICKS intersection. (Copyright Jeppesen Sanderson Inc., 1978, 1996; all rights reserved; not to be used for navigation)*

*toward the station,* or to the right in this case. (Prove it to yourself; before reaching VICKS, mentally stop the airplane and turn it to a heading of 247 degrees—you'd have to fly right to get to that radial.) If you had tuned Bradford and the needle came to rest on the side of the instrument away from the station (that is, to the left), better let Center know about it right away—you've passed VICKS.

And so to the fourth situation, the one you will use most frequently: figuring out the proper wind correction angle, or "bracketing the course." If you always take immediate corrective action when you notice the CDI drifting off center, you'll never get off course far enough to need bracketing; but a sharp turn over a VOR, or vectors to a new radial where the wind effect is unknown, may find you chasing a CDI that's off-scale.

Assuming no knowledge of the wind direction or velocity, the first thing to do is turn 30 degrees toward the radial and wait for the needle to center. For example, trying to track outbound on the 360 radial (Fig. 9-2), you find that a heading of 360 allows the airplane to drift to the right (CDI moving left). When a heading of 330 degrees puts you back on course, you have established the maximum "bracket" within which the correct heading lies—at 360, you cannot stay on course, and 330 will take you back to the radial. Now, cut the difference in half, turn to 345 degrees, and again watch the CDI. If it stays put, 345 is the correct heading. But if the needle starts to the left again, immediately turn to 330—the heading you know will return you to course—and it won't take nearly as long as the first time. Since 345 degrees has proved insufficient, take the next half step and turn to 340 degrees; be practical, accept the nearest 5-degree increment. Before long, you'll have a heading that will immobilize the CDI; this process works just as well in either direction.

This sounds like a terribly complicated, time-consuming process on paper, but in the air, it will take only a few minutes; as your experience increases, you will be able to tell roughly how much correction you'll need by observing the rate at which the CDI moves. Your target is a dead-center needle all the time, which may explain why really good instrument pilots spend so much time checking their VOR receivers; the CDI remains so motionless, they have to make sure the set hasn't failed!

# ✚ No Ident? No Good

It stands to reason that the only stations that can be used for IFR enroute operations are those that are transmitting a usable signal. The

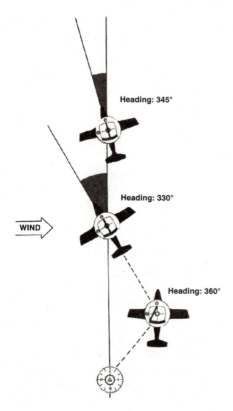

**Fig. 9-2**  *Bracketing a VOR course.*

designers provided a built-in alarm system to alert you when a VOR fails, when it is shut down for maintenance, or when the signal quality drops below a certain level. It's a very simple scheme, consisting of either automatic or manual removal of the station's Morse code identifier when any of these conditions exists. Your reaction should be equally uncomplicated: Whenever you tune a VOR and there's no identifier, *don't use it!* If an explanation of the outage isn't printed on the chart nearby, let someone in ATC know of the problem. They'll check their monitoring devices and take the appropriate action.

## ✝ Distance-Measuring Equipment

If your airplane is equipped with distance-measuring equipment (DME), you'll probably have it turned on throughout the flight, so why not use it for identifying intersections? With the CDI centered, it's merely a matter of flying along the radial until the charted number of miles shows up on the DME indicator. Establishing yourself at an intersection with the help of DME is valid *only* when you use a station directly ahead or directly behind; authorized DME intersections are shown on the Jeppesen charts by the letter D and on the government charts by an open arrowhead. (Check the chart legend for details.) The same ahead-or-behind limitation must be observed if you intend to derive accurate groundspeed readings from your distance-measuring equipment.

# ✝ VOR Accuracy Checks

The odds are probably better than even that the next trip you fly IFR will be illegal, at least from a VOR receiver standpoint. Checking VORs is one of those things that almost all pilots know about, but somehow just don't take the time to accomplish at the required intervals. According to the book, you may not fly in the IFR system unless the VOR receivers have been checked for accuracy within the preceding 30 days. Furthermore, the pilot in command is ultimately responsible for making sure this requirement has been satisfied. It matters not whether you're flying the same old airplane that you've owned for 10 years or the brand new Barnburner you rented for this one trip. You're the pilot in command, and you're the one the FAA will come looking for if they suspect a VOR malfunction is involved in a mishap.

The way to stay current is to check the receivers on every flight. The system designers have made it easy for you by putting a ground test station (VOT) on most large airports and by designating certain spots on other fields where a nearby VOR signal can be received and used for testing. If time is running out and you are far away from either of these, there are designated airborne checkpoints, so there is no excuse for flying with VORs that have not been checked. Look in the *Aeronautical Information Manual* or Jeppesen's "J-AID" for these facilities and checkpoints.

If all else fails, you may designate your own airborne check: Pick out a prominent landmark *on an airway* preferably 20 miles or more from the station and position the airplane directly overhead.

These situations should result in a centered CDI when the OBS shows the appropriate number (radial or course). Should you have to turn the OBS more than 4 degrees either way to center the needle for ground checks or more than 6 degrees either way for an airborne check, your set is out of tolerance, and IFR flight would be illegal.

Checking the VORs every time you fly is a fine habit to form, but all the effort and good intent will go for naught unless you make a record of what you have done. As soon as you have checked the receivers, make a note of the date, the location, the bearing error (how much you had to rotate the OBS to make the needle come to center), and affix your signature thereto. Where you maintain this record of VOR checks is immaterial, as long as *you* know where it is, because it's one of the things the inspectors may ask for when they investigate an IFR incident. By making some kind of a VOR check before every IFR flight, you are effectively taking yourself off the hook.

There is another way you can accomplish this check, but it must be placed in the last-resort category; if no other means is available within the time limits (and that's difficult to imagine because you can hardly fly for 30 days anywhere in the United States and not have been *someplace* where there was a VOR checkpoint), you may legally check one receiver against the other. Tuned to the same VOR, the OBS indications may not be more than 4 degrees apart with CDIs centered. But what if the first VOR is already 10 degrees off? Treat this check as something you will do only if there's no other way, and at the earliest opportunity, check the receivers properly.

# ✝ The HSI: The Complete Picture at a Glance

The VOR system appeared in the 1950s and was hailed as the salvation of aerial navigation; after struggling with the Morse-code As and Ns of four-course low-frequency radio ranges, the simplicity of VOR seemed like an instrument pilot's Valhalla. Although the principle of VOR navigation hasn't changed over the years, the avionics people have provided better ways of displaying the information; there are rectilinear needles (which remain vertical as they move across the display instead of swinging back and forth like windshield wipers) and electronic displays. But even with these much-improved instruments, you still must conjure up a mental image of your position relative to the VOR station or a selected radial; and that isn't always easy, especially when you're headed in the wrong direction or when a strong wind has drifted you way off course in a holding pattern.

The solution is a pictorial display of the navigational situation, known as a Horizontal Situation Indicator (HSI) or Pictorial Navigation Indicator (PNI). By either name, it's the greatest thing since sliced bread for *any* instrument pilot, and especially for beginners. Why not learn VOR navigation the easy way from the very start?

In an earlier discussion of attitude instrument flying, the artificial horizon was referred to as the "heart of the system," a well-deserved definition because the AI shows two aircraft attitudes—pitch and roll—on one instrument. The HSI is the heart of a pictorial navigation system for the same reason: It provides more information at a glance than any other navigation display (except a moving map) on the instrument panel. A typical installation (Fig. 9-3) shows magnetic heading, desired course, TO-FROM indication, and perhaps most important, an airplane symbol on the face of the instrument so that your position relative to a radial or a localizer course can be visualized at any time. The HSI actually superimposes VOR information on the heading display; there's no need to look in one place for the CDI, the OBS, and the TO-FROM information, in another place for heading, and then try to put them all together in your mind's eye and hope you've interpreted everything correctly. With an HSI, your mind's eye can relax because the entire navigational picture is right there in front of you. Pretend that the HSI is mounted in a hole in the cabin floor, enabling you to look down and see your airplane's relationship to a course line painted on the ground; the sooner you think of it this way, the sooner your HSI will really begin to work best for you.

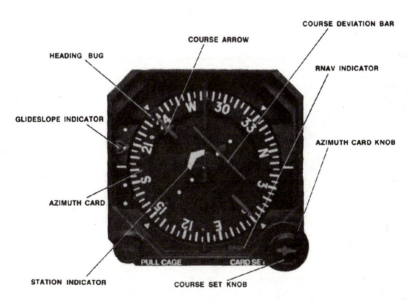

**Fig. 9-3** *Components of a typical HSI. (Courtesy of Narco Avionics)*

# Enroute Use of the HSI

Just like any other navigational device, an HSI must be properly programmed. In this respect, it's no different from the traditional VOR indicator, so your first task is to set the course selector (arrowhead) to the correct number. In the example (Fig. 9-4), a course of 120 degrees has been selected. Now the pictorial display begins to take shape; the VOR is ahead of the airplane, the course deviation bar is displaced to the right of the airplane symbol, and it becomes obvious that your present heading of 150 degrees will "fly" the airplane symbol to the radial on a 30-degree intercept angle. Unless you're flying into hurricane winds, the airplane and the course deviation bar will eventually come together, indicating that you are exactly on the desired course. Train yourself to fly the little airplane to the course deviation bar; think of the displaced segment as the desired course and fly to it. As you turn the real airplane, the azimuth ring (heading indicator) and the course selector also turn; the entire display is constantly updated.

Once you're on course, wind correction is simply a matter of keeping the airplane symbol centered on the course deviation bar by means of appropriate heading changes. And when you've turned into the wind the proper number of degrees, you'll once again have a picture of the situation—the airplane will be directly over the course deviation bar with the wind correction angle displayed as it really exists. The station pointer (TO-FROM indicator) will always indicate the location of the VOR—either ahead or behind—in relation to the selected course.

**Fig. 9-4** *Course intercept as displayed on an HSI. (Courtesy of Narco Avionics)*

Even the best of the HSIs will develop a nervous needle when the aircraft approaches the VOR station; that's a characteristic of the transmitter, not the airborne equipment. But the heading that got you to the station on course will carry you through the confused-signal area and out the other side; so when the course deviation bar starts its close-to-the-station shuffle, hold what you've got. Station passage is official when the TO-FROM indicator makes a complete reversal. Outbound tracking is no different than inbound, except that the pointer shows the station behind you.

To illustrate another of the frequently used pictorial features of an HSI, consider an airway that changes course significantly at a VOR. Suppose that, by the time station passage is confirmed and the new course is set up, the course deviation bar has moved nearly full scale to the right; a right turn is indicated, but how many degrees? What heading will result in a course intercept within a reasonable time and distance?

Until now, the intercept solution has required some computation, some guesswork, or most likely, a hit-and-miss combination of the two. But the HSI changes all that to a positive navigational technique that works when inbound or outbound; reset the course arrow to the new number and then turn the airplane until the lubber line at the top of the display touches the offset portion of the course deviation bar. (Some HSIs don't have lubber lines quite long enough to do this, but you can line up these two components of the display by mentally extending the lubber line to the tip of the offset segment.) This simple procedure puts the airplane on a positive intercept heading (usually 45 degrees), and the display shows precisely where you are in relation to the desired course. Sooner or later, the deviation bar will begin to move from its fully displaced position; when this happens, turn the airplane as required to keep the lubber line and the tip of the offset bar together until the intercept is completed. This technique is smooth, positive, and *pictorial*; it makes the tendency to overshoot radials a problem of the past.

## Holding with an HSI

Let's face a fact of IFR life: A holding pattern entry flown with a nonpictorial VOR display has an element of confusion built in because you can't readily see where you are in relation to the holding course. On the other hand, an HSI makes holding patterns understandable. Regardless of the type of entry you elect to use, set the course arrow to the *inbound course*—the holding course—when you pass over the station the first time. The course deviation bar will immediately display the position of the all-important holding course, and the airplane symbol will erase all doubt about where you are. When the first minute is up, merely turn the airplane back toward the holding course, complete the intercept as described earlier, and go about your business with confidence. There's no problem with so-called reverse sensing here; when the course arrow is set to a desired inbound course, the HSI will *always* display the location of that course in its correct relation to your airplane.

## HSI and ILS: Partners in Precision

All the good things about an ILS procedure are made better when there's an HSI in your instrument panel. While it's no more accurate than any other display, it enables you to see where you are throughout the entire approach; and regardless of the amount of wind correction required to stay on the localizer, the course arrow represents the landing runway and, in a strong crosswind, indicates how far left or right you'll need to look for the approach lights.

There's a cardinal rule to remember when setting up the HSI for an ILS or localizer approach: *Always set the course arrow on the published front course.* This rule is true for *all* localizer work, even when flying a back-course approach. Set     up this way, the course deviation bar will always display the localizer, and your position will be apparent in terms of the airplane symbol's relation to the offset segment of the bar.

Whether vectored to the final approach course (the most common situation these days) or completing a procedure turn, the localizer intercept should be accomplished the same as turning onto a VOR radial; keep the lubber line directly over the tip of the course deviation bar, and you'll arrive on the localizer automatically, with no overshoot.

Having intercepted the localizer and proceeded toward the marker, the HSI should be flown exactly as you would fly the "normal" presentation—when the bar slides left, turn left, and vice versa—but with the added safety and convenience of obtaining all the essential navigational information from one source.

The HSI's glide slope indicator is interpreted the same as any other; it's always directional and always displays the location of the glide path relative to your airplane. When it moves above center, you should decrease your rate of descent; when it's on the mark, you are on the glide slope; when the indicator slips below center, you should power down and descend a little faster to regain the proper position.

A back-course localizer approach, that notorious boggler of pilots' minds, becomes crystal clear with an HSI. All you need remember is to set the *head* of the course arrow on the front course of the ILS; the offset segment of the display now shows the localizer in its true relation to the airplane symbol no matter which direction you fly. Thus, a back-course approach literally becomes as easy to fly as the front course, with absolutely no concern about flying "away from the needle." When properly set up for enroute, holding, or approach, the HSI's pictorial display reduces the navigational task significantly; just fly the little airplane toward the deviation bar to resolve any off-course situation.

## Other Ways to Use an HSI

Even when there's no signal being received and displayed, an HSI can help you visualize the aircraft's position. As the airplane turns, the azimuth ring turns, and the course arrow always shows you the relation between aircraft heading and desired course.

For example, when setting up for an NDB approach, put the HSI course selector on the *inbound course* published for the approach. As you pass over the beacon and proceed outbound, notice that the two arrowheads—one on the ADF pointer, the other on the HSI—are pointing to the same place (the tail) on their respective dials. At the proper time, turn left or right 45 degrees to start the procedure turn. Here, once again, both arrowheads indicate the same 45 degrees from the tail position. (The HSI has convenient 45-degree "tic marks" all around the dial, so you no longer need to look at or compute procedure turn headings.) The 180-degree turn puts you on a 45-degree intercept, and the HSI's pointer will stop exactly 45 degrees from the aircraft heading. When the ADF needle is parallel to the HSI course arrow (both indicating 45 degrees from the nose of the respective airplane symbols), you're on course—it's just that simple. By always setting the HSI arrowhead on the *inbound course,* the relation between the two arrows is consistent, and all the computation from an NDB procedure is eliminated. This technique works just as handily when holding on an NDB.

You can also use an HSI to help orient yourself on the approach to a completely unfamiliar airport. When the course arrow is set to the number that represents the landing runway the HSI displays a picture that can hardly be misinterpreted; you'll "see" the runway, and can plan which way to turn and how much when the field comes into sight.

On a large airport with several runways and taxiways, set the HSI to the runway heading before taxiing; at least, you'll be able to make a sensible determination of whether to turn left or right as you leave the ramp.

In summary, an HSI would cause that ancient Chinese philosopher to update his comment about the value of a picture: "One HSI is worth a thousand left-right needles."

# 10

# Area Navigation (RNAV)

Crows are not the bird world's fastest flyers, but in the process of exercising their legendary ability to get where they're going in straight lines, they're bound to save some time when flying from tree A to tree B. Instrument pilots anxious to maximize the efficiency of their airplanes should also think in terms of straight lines, but it seems impossible when our primary navigational system (the VOR, or "Victor" airways) is built on IFR routes that dogleg their way across the country. Every now and then you'll come across a situation in which there's a VOR at the departure airport and another at destination. Then an accurate, straight-line course is possible, but that's the exception, hardly the rule.

On the other hand, there are a thousand or so VORs scattered across the United States today, and almost anyplace you'd want to go in an airplane is on some radial of some VOR; however, you must be on that radial to know where you are. Plotting two radials that cross above the destination is a big help, but it's still a blind navigational situation until you locate yourself on one of the radials and fly until the other needle centers.

The advent of the VORTAC (TAC refers to the military's Tactical Air Navigation system and indicates DME capability for civilian users) and the self-explanatory VOR-DME facility mean that any point within the service volume of that station can be defined in terms of a radial and distance. For example, you might be located at a point 15 miles from the VORTAC on the 225 degree radial and the place you'd like to go might be on the 315 degree radial, 15 miles out (Fig. 10-1). Any student who managed to stay awake in Trigonometry 101 would recognize the triangle thus produced and could offer a solution. When the

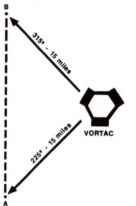

**Fig. 10-1**
*The RNAV triangle.*

lengths and angular relationships of two sides of a triangle are known, a simple calculation can be used to figure out the length of the remaining side and, in this case, the direction of the third side with respect to magnetic north.

But few pilots have the time (or the inclination) to consult a table of trigonometric functions and come up with the answers, especially when the airplane has moved several miles during the calculations. Why not give the job to a computer that has been programmed to figure out these triangles and come up with the right answers continuously and instantaneously? When the electronics people were able to reduce computers from the size of your living room to the size of a matchbox, area navigation (RNAV, for short) became possible. Now, instead of having to fly the dogleg airways and fumble with two-VOR fixes, you can navigate directly to or from any point within the area served by a VOR/DME facility. RNAV opens a whole new world of efficient, straight-line navigation. Look out, crows!

RNAV has introduced a new term—*waypoint*—to the aviator's vocabulary. This is the proper way of referring to those locations defined by radial and distance from a particular VORTAC. Because a waypoint can be placed anywhere you want it—on an airport, a Final Approach Fix, a Missed Approach Point, or a mountain (so you can stay away from it)—RNAV makes possible a navigation system with complete flexibility that allows you to choose your own route.

A typical RNAV installation consists of an onboard computer, a means of entering the "waypoint address" (that's RNAV talk for the radial-distance information that defines the waypoint), and a display that shows magnetic bearing and miles from the selected waypoint. An electronic interface sends the directional signal from the computer to the Course Deviation Indicator to provide steering information. From an operational standpoint, you fly RNAV exactly the same way you fly a normal VOR, but the LEFT-RIGHT and TO-FROM indications, the setting of the OBS, and the computer-generated range are based on a waypoint instead of a VOR station.

Nearly all RNAV manufacturers have taken advantage of the microprocessor revolution to provide additional navigational data at the touch of a button. For example, you can call up the radial and distance from the VORTAC you're using—"raw data" (uncomputed)—and that information comes in handy to satisfy your occasional curiosity as to your location, to say nothing of the easy, accurate reply to ATC when a controller asks, "Where are you?" Pushing another RNAV button will display groundspeed to or from the waypoint and the time required to fly there.

You'll not fly with RNAV very long without noticing that the CDI has lost its "Nervous Nellie" characteristic close to the station and that RNAV station passage is accompanied by a "soft" change of the TO-FROM indicator. These welcome changes are due in part to electronic filtering as the raw data are massaged by the computer, but more important, the steering display (CDI) is based on a totally different and more realistic concept than conventional VOR: RNAV displays *linear deviation.*

On course (CDI centered), RNAV and VOR presentations are identical, but the picture changes remarkably when your airplane moves to the left or right of a selected course. The VOR-only installation shows how many *degrees* you are from the selected course; RNAV presentations indicate how many *miles* you are from the desired course. VOR indicators are designed with a scale of 10 degrees either side of center, whereas most RNAVs have a scale of 5 miles left and right in the enroute mode.

To illustrate the difference, imagine your airplane equipped with both standard VOR and RNAV, both set up on the same station/ waypoint and with the same course selected on the OBS. When the airplane is 2.5 *degrees* off course, the VOR indicator will show one-quarter of the 10-degree distance between center and full-scale deflection; if that same location is also 2.5 *miles* off course, the RNAV indicator will show a displacement of one-half scale.

As you continue flying straight ahead in this situation, the VOR needle will show more and more deviation, and since the angle is constantly increasing, you are flying farther and farther from the selected radial. On the other hand, the RNAV CDI sits there steady as a rock, telling you that the course you have selected is—and remains—2.5 miles to one side. You are flying parallel to the desired course, something that's impossible with ordinary VOR. In both cases, the TO-FROM indicator changes as you fly by the station/waypoint, but with RNAV, you'll always know the displacement *distance*. This can be a very useful feature, such as when you'd like to arrive on a 2-mile downwind leg at a strange airport. Set up a waypoint on the airport, fly a parallel course on the desired side with the CDI 2 miles off center, and when the TO-FROM changes, you'll be at midfield on a 2-mile downwind.

From the very beginning, RNAV units have incorporated a feature that lets you select a more sensitive CDI mode for RNAV approaches and other situations that require a finer display of off-course distance. Commonly called the "approach" mode, this selection changes the scale of the CDI; typically, the CDI now indicates only 2.5 miles either side of center, and small deviations from course will be easier to see. In general, the approach mode should be selected when you're approximately 20 miles from the waypoint and should remain in that configuration throughout an RNAV approach procedure.

## ✝ Putting RNAV to Work

One application of RNAV is for relatively short trips from and to your home base. With a VORTAC located close by—or at least within reception distance after you've climbed 1000 feet or so—you can begin building a library of waypoint addresses for oft-used airports, based on the radial and distance from that nearby VORTAC. Within the RNAV equipment's range, you'll then have positive course guidance and distance information at takeoff or very shortly thereafter.

For the return portion of a trip, you might draw a circle around your home airport, the radius of the circle being the working range of your RNAV equipment. Choose VORTACs that lie generally north, east, south, and west of your airport on or inside the ring and plot the waypoint addresses for your airport from these stations. Now, when you're inbound and within range of one of the preselected VORTACs, load the home-base waypoint, confirm that you're getting a usable signal, and ask ATC for a clearance *direct*.

RNAV flight planning involves little more than drawing a course line from departure to destination and selecting waypoints at convenient intervals along the course. Make it easy by selecting waypoints on cardinal radials (090, 180, 270, 360), and you have to change only the distance when loading the next address. Once under way, navigation is just a matter of flying from one waypoint to the next, checking carefully to be certain that both VOR and DME units are properly tuned, that the stations are positively identified, and that the waypoint address is correct.

## ✝ Filing an RNAV IFR Flight Plan (Using VOR-DME RNAV)

The suffix R after the aircraft type in an IFR flight plan indicates that, in addition to an altitude-encoding transponder, the aircraft is equipped with an FAA-approved RNAV system. Beyond that, your flight plan for an RNAV trip is no different from one using VOR airways except for the language used to describe the route of flight. Let the specialist know that it's an RNAV flight plan and then go ahead with the waypoints you intend to use; for example, "direct LMN 270026, DSM 270022, FOD 090030, direct MCW." In addition to waypoints selected for your convenience, include one for each turning point (if any) on your route.

Make certain your route of flight includes at least one waypoint in each Center area you will pass through and that such waypoints are located no more than 200 miles from the border of the previous Center's bailiwick. You should also be aware that one of the prerequisites for ATC's acceptance of a straight-line RNAV route is the capability of radar following throughout the flight; this may be the reason an RNAV-direct request is sometimes turned down.

The final waypoint of a typical RNAV trip is right on the airport, but it doesn't meet the standard of good practice, which calls for the final item in "Route of Flight" to be the radio fix from which you intend to commence an instrument approach in the event of communications loss. If the destination airport has a published RNAV approach procedure, it would be proper to list the appropriate Initial Approach Fix as the last item. This could be an IAF shown on the RNAV approach chart or a non-RNAV fix on a conventional chart.

## ✝ RNAV Includes LORAN and GPS

The simplicity of planning, filing, and flying IFR with RNAV rises to new heights for pilots with long-range navigation equipment. The two most prominent these days are Loran (LOng RAnge Navigation) and GPS (Global Positioning System). If we can believe what we hear from the government, GPS will eventually replace Loran and the existing VOR system, but don't hold your breath; the next time a new navigation system doesn't come on line when it's scheduled won't be the first time.

Loran, which now covers almost all of the continental United States, works its magic by interpreting the time difference in low-frequency radio signals from chains of master and slave stations. A typical Loran display shows present position, course to be flown, course deviation, groundspeed, ETE, etc. More accurate than VOR-DME RNAV but not as precise as GPS, perhaps the best thing about Loran is the fact that the signals can be received almost anywhere, including on the ground and behind mountains where VOR signals are blocked.

GPS operates on signals transmitted from a number of orbiting satellites and provides virtual worldwide coverage with accuracy unprecedented in aerial navigation. The cockpit display is generally similar to that provided by Loran.

Remarkably different as they are in terms of operating principles, Loran and GPS both provide course guidance from one point to another along a "great circle" route, which is the shortest distance between two points on a sphere. Imagine a world globe with an equator that can be moved about until it rests on your point A and point B; that portion of the equator between A and B is a "great circle" route.

So in accordance with good navigational practice, you should draw a course line on the chart to help keep track of your position in relation to alternate airports, restricted areas, terrain, obstructions, and so forth as you go on your way. But as far as ATC is concerned, only Point A (departure) and Point B (destination) are important, and the computers in the Air Route Traffic Control Centers are programmed to recognize and accept great circle tracks. You should specify turning points on your route and include a checkpoint within 200 miles of each Center's boundary. If the destination airport's library of charts includes STARs, or if you are aware of a specific approach fix in use, file to the appropriate point instead of to the airport itself.

Obviously, there are times when other traffic, hot restricted areas or MOAs, or local flow management procedures prevent a controller from accommodating your request to fly direct, but it's worth a try. If your "RNAV direct" request is turned down before takeoff, try again when you're in the air. Each time you are handed off to a new controller, politely request "RNAV direct." It's not at all uncommon for the traffic situation to be completely different as you get closer to destination, and such a request is often granted. Why not try to wring every bit of utility and economy out of that expensive navigation equipment?

As with any off-airways journey, you are responsible for selecting and accepting an IFR altitude that guarantees proper obstacle clearance throughout the flight. With a Sectional, a WAC chart, or an RNAV planning chart, check terrain and obstacles along the way and be certain that your flight altitude clears everything within 4 miles either side of course by at least 1000 feet in the flatlands and 2000 feet in those areas designated "mountainous."

# ✝ RNAV (Primarily GPS) Approach Procedures

About 30 years ago, it looked like VOR-DME RNAV was the wave of the future, and published RNAV approach procedures brought IFR capability to some airports that were otherwise limited to VFR operations. Many of these procedures still exist, but the now inevitable demise of Loran, the probability of an eventual VOR phaseout, and the proliferation of GPS equipment in the general aviation fleet have resulted in the emergence of GPS as the primary nonprecision approach system. You've no doubt noticed that new approaches (see Fig. 10-2) are added to your chart books on a regular basis, and there's no reason to think it won't continue. Sooner or later, technology will enable GPS to provide a glide slope as well as accurate course guidance at virtually any airport—and there goes the ILS as we know it.

But for those pilots who continue to use VOR-DME RNAV, understand that a published RNAV approach is nothing more than another nonprecision procedure using a VOR-type display for course guidance. The procedure, published in chart form, will be familiar to any instrument pilot, the only major change being the use of waypoints instead of VORs, NDBs, crossing radials, and the like. You will often need help from an approach controller because some of the IAFs in some RNAV procedures stand alone in the airspace with no feeder routes, procedure turns, or lead-in radials.

Some conventional RNAV equipment enables you to select a "cross-track" or "parallel-track" mode, in which case you can fly with the CDI centered and be assured of a specific offset from the desired course. This feature can work to your benefit when conducting a circling approach or setting yourself up for a properly spaced downwind leg in VFR conditions. Get to know the capabilities of your RNAV equipment and use them to your advantage.

Good VOR-DME RNAV equipment is quite good. You'll be pleased with its accuracy and delighted with the integrity of the LEFT-RIGHT display. Because it is driven by computed data, the CDI doesn't respond to many of the electronic glitches that plague raw VOR data and make for a nervous needle. But just because it's better than plain vanilla VOR, don't allow yourself to become so complacent and over-confident that you press on below the published RNAV MDA or into visibility conditions that are well below the requirement on the approach chart. An unscheduled contact with the ground is just as unpleasant during an RNAV approach as any other.

GPS is all this and more. It's considerably more accurate than conventional RNAV, and although neither of these systems have much effect (if any) on the MDA for a VOR or NDB approach, you'll be happier with the increased accuracy of GPS. When you break out with the CDI centered, you should be on the runway centerline.

Most GPS equipment has the added convenience of "automatic" approaches. That is, the black box will line up the approach fixes in the proper order, make the changeover as each waypoint is passed, and let you know what it's doing. Marvelous stuff, this, but don't get so engrossed watching the electronic wizardry that you descend below MDA sooner than you should or forget to look for the runway or other airplanes. Black boxes are indeed wonderful, but they are not a substitute for good judgment, common sense, and situational awareness.

# ☦ Other Applications of RNAV

One of the nicest things to know on a nonprecision approach is your distance from the Final Approach Fix and, subsequently, how far you are from the airport itself. With RNAV equipment, the distance question is answered with considerable accuracy; you can put a waypoint on the FAF, another on the airport, and even if course guidance must officially be obtained from another source, accurate range information is a great help. (You can use RNAV steering information to back up the primary navaid, but unless it's a published RNAV procedure, that's the best you can do.) This is especially true in the case of an on-airport radio aid, which previously required an extension of the time outbound and occasionally a "plunge" after the final approach

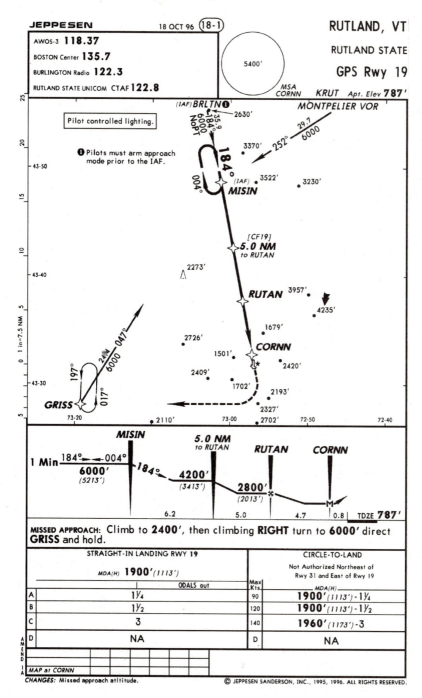

Fig. 10-2   *Typical GPS approach procedure chart. (Copyright Jeppesen Sanderson Inc.; not to be used for navigation; all rights reserved)*

fix to guarantee reaching MDA prior to reaching the airfield; with an RNAV distance display working for you, all the guesswork is gone.

There's not a pilot flying today who wouldn't agree that always having an alternate airport in mind—at *any* point on *any* trip—is a good idea. The assurance of someplace to go if things start to go bad is especially important at night, in weather, or both. Enroute with RNAV, you can keep inactive waypoints programmed for "instant alternates" throughout the flight; it takes a little extra preflight planning, but being able to push a button in time of need and have an accurate course and distance to a safe haven come up on the cockpit display is luxury. Upgrade that to *sheer* luxury when you have a contemporary Loran or GPS in the panel because these units are preprogrammed with nearly all the airports in the country, and they display the nearest airport information automatically.

Holding at an airway intersection or an NDB is a piece of cake with RNAV. Instead of navigating on one radial while watching for the other to center, figuring out what the DME reading should be, or tying yourself in navigational knots trying to kill wind drift over an NDB, you can put a waypoint on the holding fix and fly on the proper radial until FROM appears; area navigation makes *every* holding fix a VOR.

Or how about this one? You're inbound to a busy terminal, visibility barely 3 miles in summer haze, and you are asked to report on a 3-mile final. Up until now, that position report has been little more than a good guess—especially at a strange airport—but with a waypoint on the airport, you can line up on a genuine final approach (set the OBS to the runway heading and keep the needle centered) and let the controller know when you're *exactly* 3 miles out; that's a big help for everybody concerned. And if you want to show off a little, get *super*accurate—tell the tower that you're "3.1 miles on final."

## ✝ Summary

Most nonprofessional instrument pilots can probably be categorized as those who fly only short IFR trips close to home for currency or practice, or those who use their ratings for frequent, relatively long IFR flights. The former group might as well save the money they'd spend for RNAV equipment—at least until the plug is pulled on the VOR system—because there's lots of quality instrument time to be logged using the federal airways and VOR approaches. But IFR "frequent flyers" should give serious consideration to equipping their airplanes with Loran or GPS because of the significant benefits these black boxes provide. (If you're flying with a good Loran set, hang on to it. Not only will it serve you well until the government phases out the system, but by the time you are forced to make the switch to GPS, the technology will have improved remarkably. The price will likely be lower as well.)

There's a lot of navigational information available from even the most unsophisticated Loran installation, and when it comes to GPS, well, the stuff you can bring up on the screen is truly mind-boggling. As you become familiar and facile with long-range equipment, you'll find more ways to use that information to improve the flexibility and efficiency of your IFR operations.

But perhaps the most significant advantage for Loran and GPS users who fly frequently and far is the ability to simplify their flight planning and their inflight navigational chores by filing and flying *direct*. It was mentioned earlier in this chapter, but based on personal experience, you should file RNAV direct from airport to airport at every opportunity—quite literally, every flight. You'll find that your request will be turned down only on rare occasion, and even when that happens, it's highly probable that you can pick up a clearance direct to your destination at some point along the route. For the Loran/GPS crowd, the VOR system exists as a backup to both enroute and terminal IFR navigation.

We're in the midst of a near explosion of avionics technology, and this time, small airplanes are not being shut out. It's a lot like the need to get on the information superhighway if you're going to remain competitive in whatever you do—including flying IFR—so you should equip yourself and your airplane accordingly for the future.

# 11

# Enroute Procedures

When you're settled at your assigned altitude, when you're finished with the sometimes demanding communications of the departure phase, an IFR trip becomes largely a matter of flying from one check-point to the next. An occasional traffic advisory or sector handoff may be your only contact with Center; you may feel like calling the controller now and then just to make sure there's still somebody there. In the enroute portion of a flight, there's time to plan, check the weather ahead, and give most of your attention to flight management.

On your first few flights, especially in weather all the way, you will no doubt hover over the gauges like a mother hen, but as your experience builds, you'll discover that the airplane does a very good job of flying itself. You should be more a navigator than a pilot at this point, more a manager than a manipulator. This chapter contains techniques and suggestions that should make your enroute opera-tions less confusing, more efficient, and above all, safer. Once again, knowledge is the key a thorough understanding of what you're doing now and what you may be doing in the miles ahead will make you a better instrument pilot.

## ✛ First, Fly the Airplane

At the head of the list is aircraft control because, without it, nothing else counts. Trim is a key element, and barring rough air, you should be able to set up your airplane so that it will fly for relatively long periods of time with only gentle nudges on the controls to maintain the desired altitude and heading. Most of today's aircraft are rigged so that rudder pressure is hardly required in turns, so "feet on the floor" is not at all a bad way to fly IFR in smooth air; it will help you to relax on long flights when keeping your legs in one position for extended periods can cause a distracting amount of muscle fatigue.

If you have invested in a wing-leveler or autopilot, by all means use it when you're flying IFR; it takes a big load off your mind and lets you focus attention on more important things. (It's even better than a copilot because it won't drink your coffee or talk back!) An automatic pilot with approach-coupling capability is even better. Study the operating manual and find out what your autopilot can and can't do. Use it most of the time; you're not going to increase your instrument skills a great deal by hand-flying straight and level. At regular intervals—maybe every third or fourth flight—fly the airplane manually through the departure, climb-out, and make the approach hands-on so you won't lose the touch. Autopilots have been known to malfunction; no superstition implied, but they always seem to fail when they're needed most!

## ☩ Like a Scalded Duck

Climbing to your assigned altitude or moving to a new, higher level when requested by ATC should be done at the best rate of climb speed, with the safety-related option of climbing at a higher airspeed in VFR conditions so that you can see better over the nose. Your interest on the initial climb should be to get up there as rapidly as practical, out of the low-level turbulence, out of the icing layers, and into clear air whenever possible. Your flight planning was no doubt based on the selection of an altitude as high as practical for you, the airplane, and the existing weather conditions, so get there as soon as you can. Cruise airspeed will inevitably be higher than that used for climb; because time is of the essence, spend as little of it as possible at low airspeed.

## ☩ Trade Altitude for Airspeed

The reduced-power, constant-airspeed descent you use for proficiency maneuvers should be reserved for your practice sessions under the hood. When descending in the real world, the potential energy stored in the airplane as a result of climbing to altitude can be converted to airspeed on the way down, and that pays off in time saved. The limitations to be observed are airspeed and "ear speed," the former a function of maximum speeds for the airplane structure, the latter concerned with the rate at which you and your passengers can clear your ears. There's absolutely nothing wrong with descending at a high airspeed, as long as you stay under the never-exceed speed and carry enough power to keep the engines warm. (Of course, if you encounter anything more than light turbulence during the descent, you've no choice but to stay at or below maneuvering speed—corrugated wings went out with Ford Trimotors.)

Advise your passengers that "we're going to a lower altitude"; never use the phrase "we're going down," as some folks interpret this as advance warning of a crash and will try to get out of the airplane. Remind them to keep swallowing or yawning as you descend. It's far better to put up with a grumbling passenger or two who resent being awakened for a descent than to have them cursing you and aviation in general for the pain in their ears the rest of the day. Most people can tolerate high rates of descent if they know what's happening and what to do about it; if you have doubts, take it easy.

Most supercharged engines have manifold-pressure controllers built in, and you can come down through increasing atmospheric pressure without much concern; but normally aspirated engines are something else. Know the maximum manifold pressure for the rpm you have selected (or the limiting engine speed for fixed-pitch props) and let the pressure or rpm build up to that point; then make small throttle reductions to keep it there. You'll have longer-lasting engines and will still be able to realize the high airspeeds that cut precious minutes from the total time. (Refer to the power chart in your aircraft handbook for maximum settings at various altitudes.)

When you have determined a comfortable rate of descent (experiment until you find the one that suits you best), divide it into the altitude that needs to be lost; this will give you the *time* required to descend. When you make an estimate of your groundspeed in the letdown, you can compute a *distance* from destination at which to request a descent. In a crowded terminal area, this may be impossible because of other traffic, but plan ahead and ask; you've nothing to lose but time.

## ✢ Enroute Charts Make Good Flight Logs

Your IFR route will usually be just what you requested, and maintaining a flight log is good practice; but that involves another piece of paper in the cockpit, and the most carefully planned route of flight turns useless when ATC throws a change at you. Use a soft, easily erased pencil to write your actual times of arrival and the estimates for subsequent fixes on the enroute chart. That puts the information right in front of you, available for quick reference (Fig. 11-1). Put ATAs below the fix, ETAs above. (Mark down the times when you switch fuel tanks the same way.)

"In radar contact" is music to an instrument pilot's ears because it does away with position reporting. But it breeds a complacency that lurks in the shadows, waiting for the time when radar contact is lost and you have to resume reporting. This means figuring ETAs, keeping some kind of a log, and making all the navigational decisions yourself. Any system of aerial navigation requires that you know where you are now, where you have been, and at what rate you are moving across the ground. To forestall utter confusion, make it a habit to record the time you pass over navaids enroute. When the day of reckoning arrives, you'll have some idea of *when* you were *where*; it's a place to start.

## ✝ Pilot, Know Thy Fuel System!

The subject of fuel management on any flight deserves your careful attention, but it's more important when you're IFR because there's not

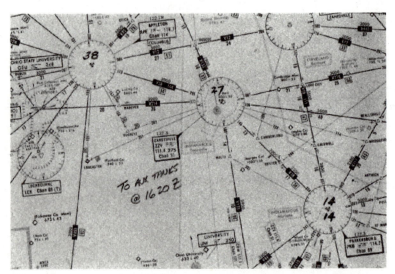

**Fig. 11-1.** *The enroute chart makes a good flight log and a handy place to keep a record of fuel management.*

always an airport handy where you can land and top off the tanks. Every airplane will be different in this respect, but two general rules apply to *all* flying machines: First, know what the range capabilities and fuel-burn rates are at various power settings; second, fly by the clock, not by the indications on the fuel gauges. When you know the amount of fuel on board at the start and keep track of how much you burn, there should be no excuse for arriving at that frightening circumstance of not having enough fuel for the distance you have to go.

Running the tanks dry is not recommended as a continuous practice (passengers don't like to be awakened by a snorting, surging engine), but you should do it at least once for each tank under the proper conditions, keeping tab on the clock so that you know *exactly* how long each tank will keep things going at cruise power. You may be surprised (pleasantly or otherwise) to discover the differences that exist between tanks; it can be 20 minutes or more on some models, and that knowledge might save your neck someday! But remember that whatever finds its way past the fuel filter (dirt, water, bits of fuel cell sealant) will likely be drawn through the engine when you try to run it on the fumes. It takes only a speck of dirt or a crystal of ice lodged firmly in a fuel line or carburetor port or injector nozzle to shut off the power completely, so you might want to limit your "dry tank" checks to conditions that will allow you to safely make like a glider.

## ✝ Stand by for the Latest Change

The routing and altitude specified in your clearance are not sacred, and ATC will frequently make changes and amendments to suit the traffic situation as you proceed. It's possible that you may be asked to fly a different route, another altitude, or hold until a traffic conflict is resolved. You should treat any such change as a *suggestion* (just as you did when copying your clearance on the ground) and refuse *any* clearance that will fly you into trouble. This is not the sort of thing that should keep you awake at night, but remember that *you* are responsible for the safety of the flight, not the controller. Don't be unreasonable about it, but if Center asks you to fly a route or altitude that you don't like, explain the situation and offer an alternative. If that doesn't work and a controller insists, exercise your authority, fly as you think you should to stay out of trouble, and be prepared to justify your actions. (Tell the controller what you intend to do, of course.)

Most ATC instructions are right to the point, with no indication from the controller that what's coming up is anything but routine. But you can almost always tell when a route change or a hold is coming your way because the controller will ask, "34 Alpha, I have an amendment to your clearance—ready to copy?" Your reply should be "stand by," which gives you time to break out the enroute chart so you can see if the upcoming change is compatible with the weather situation, terrain clearance, fuel remaining, and so on. Having consulted the chart, you can tell the controller to "go ahead with the clearance" and visualize the amendment to determine if you can comply.

## ✝ See How the Other Half Lives

If you haven't taken the time to visit an Air Route Traffic Control Center, by all means do so at your earliest opportunity. In addition to gaining an appreciation of the controllers' task, you will also come away with a firm resolve to never again put all your eggs in one basket. You will forevermore pay close attention to what ATC is doing with you *and* with other IFR flights in your immediate vicinity. ATC is *people,* and as such, they are subject to the human errors that have plagued us all in the past, which get us into trouble today, and which will continue to cause problems in the future.

Sharp instrument pilots know what's going on around them. Witness a situation several years ago in which a pilot felt that Approach Control was issuing vectors to the wrong airplane; after several heading changes that just didn't seem right, he said to Approach Control, "You're not sure *which* airplane you're looking at, are you?" A long, embarrassed silence followed as the approach controller tried to sort out the targets. In this case, everything worked out all right, but whenever you have that nagging doubt, check on it. A case of mistaken identity has no place in an IFR environment; it happens *very* infrequently—but then, it only takes once.

## ✝ A Place for Everything

Be a good housekeeper whenever you are flying, and really work at it when you're IFR. Once beyond the terminal area, put away the SID or the area chart (put it where it belongs, not on the cockpit floor) and get out the enroute chart, folded to show the route you're flying.

All the charts have information in the margins that indicates the frequencies, airway numbers, courses, and distances to on-the-next-chart facilities. When you fly off one chart, put it away and get out the next one. As you approach the terminal area, put the enroute charts back in the book and break out the paperwork you'll need to fly the approach. Find a handy spot for your navigational computer—out of the way, but always in the same place when you need it. A boom mike and a control-wheel microphone switch will pay for themselves many times over in convenience and composure. (You can buy rigs that are truly portable and can be used in any airplane you might fly.)

Your own personal shortcuts will develop as you gain experience in the IFR system, and when you've done a good job of thinking ahead and staying ahead, you may find yourself with nothing to do. But when you're tempted to sit on your thumbs, look around; there's probably something that needs to be done, or at least thought about. Recheck the fuel situation, find out what the weather is doing up ahead, or make a correction to get the CDI on dead center. The more frequently you take inventory, the smaller your corrections and adjustments will be, and this goes a long way toward making the enroute portion of your flight smoother, more enjoyable, and safer for all concerned.

# ✢ IFR in Thunderstorm Country

"Each pilot in command shall, before beginning a flight, familiarize himself with *all available information* concerning that flight. This information *must* include, for an IFR flight, weather reports and forecasts and alternatives available if the planned flight cannot be completed." That's the law according to paragraph 91.103 of the FARs, shortened somewhat and with emphasis added to make this point; when a pilot obtains "all available information," has formulated a plan B, and is ready to use it if the original plan goes down the drain, *there is no excuse for blundering into a thunderstorm.*

Despite that, some aviators continue to disregard all the signs, fly themselves into horrible weather conditions, and then fuss and fume because the controllers didn't warn them or vectored them right into the middle of a thunderstorm area. IFR pilots who are interested in living to a ripe old age must understand that *they cannot depend on ATC for weather information*; note carefully the use of the word "depend" because there's not a controller around who wouldn't help a pilot in time of need. But controllers are limited in at least two ways: the physical capabilities of their equipment and duty priorities.

In the first case, today's ATC radar is designed for controlling air traffic, not looking for thunderstorms. Not too many years ago, you could call up Center and get a good verbal description of the actual weather returns on radar; today, controllers look at a computer-generated radar image, which compromises some weather-observing capabilities for the sake of better aircraft separation.

The second limitation, that of controllers' duty priorities, is clear-cut and easy to understand:

- First priority is separation of aircraft and the issuance of radar safety advisories (low-altitude alerts and conflict alerts).

- Second priority is to provide other services that are required but do not involve separation of aircraft.

- Third priority is to provide additional services to the extent possible.

When weather information is not volunteered, pilots must assume that the controllers are busy with other duties. The message comes through loud and clear: *Pilots must not rely on ATC to keep them out of thunderstorms.* There, I've said it again.

The best way to stay safe is to avoid thunderstorms using the best detector of all: the human eye. This may mean flying low enough to see under and around the problem or climbing until you're able to circumnavigate the buildups in the clear air above. It may also mean beating a hasty retreat—the ever-popular 180—when it appears your present course will take you into clouds that might hide thunderstorms. If you find yourself flying "on the gauges" in an area with thunderstorm potential, you're in the worst possible situation. Don't fly a minute longer toward possible disaster; if ATC won't give you an immediate clearance to go somewhere else, exercise your authority and *do it!* Remember that the controllers' first priority is traffic separation; have faith that they'll work something out if you elect to fly in a different direction.

There's another implication in FAR 91.103 that is all too often overlooked: the meaning of the phrase "all available information." Your weather-checking responsibility doesn't end with the preflight briefing; it continues throughout the flight and includes listening to what other pilots on the frequency have to say about the weather conditions they're encountering, listening for SIGMETs and warnings of hazardous weather on HIWAS VORs, and perhaps most important, *asking* for current weather from anyone who might know more about the situation than you do. Good sources of weather information would include the Flight Watch stations (only a radio call away on 122.0), a Flight Service Station that you know has weather radar (that includes just about all of them today), a controller whose radar might have the information you need, or another pilot along your route of flight. There's a lot of weather news available—sometimes good, sometimes bad, depending on your point of view—but it's not all served up voluntarily. When in doubt, ask.

That first instrument flight all by yourself is a heady experience, the culmination of a great deal of expense and effort, perhaps an opportunity to begin realizing the benefits of utility and efficiency that may have started you down the IFR road in the first place.

That may have been the case with one newly minted instrument pilot who needed to get home after a day's business on the other side of the mountains. A cold front with attendant short lines and clusters of storms lay across the route of flight; consistent with his preflight briefing, the lightning flashes in the darkened sky provided positive evidence of the weather's location as the pilot proceeded westbound in his Cessna 210.

He accepted storm-avoidance vectors from Center controllers, who made it very clear that a deviation well to the south of course would be required to get around all the storms showing on radar. About 100 miles from home, the pilot inquired into the feasibility of "going direct" to his destination (which would have taken him across the line of storms), was advised that several aircraft had done just that, and that a heading of 330 degrees would take him through a light spot on Center radar.

Shortly after accepting the vector, the pilot's reports of close-aboard lightning flashes and turbulence increased rapidly, and within a few minutes, his virginal IFR trip ended in disaster. The pilot crashed in the mountains, a victim of upset and disorientation.

A postaccident investigation of the radar images did indeed .show a "light spot" on the controller's display, but the ATC radar *didn't* show what turned out to be a cyclical rise and fall pattern in this particular line of storms. The pilot ventured into a "soft spot" just before it turned into a very *hard* spot.

Lesson to be learned? Don't ever rely on ATC radar to select a penetration course through an organized system of thunderstorms. The very best you should expect of a controller in this situation is a vector to *avoid* the weather, which is precisely what ATC intended for this pilot. His decision to fly direct—even with his own visual sightings of lightning—will never be completely understood.

There's *always* a better way to get someplace in an airplane other than through a thunderstorm. The risk is never worth the price that might have to be paid.

# ✝ Using Airborne Weather Detectors

Here, in order of importance, are the airborne weather-detection devices available to contemporary pilots:

1. The human eye
2. Passive storm detectors
3. Weather radar

And here, in order of importance, are the airborne weather-detection devices available to contemporary pilots:

1. The human eye
2. Weather radar
3. Passive storm detectors

Do you get the point? For all the reasons mentioned a few paragraphs earlier, visual detection of thunderstorms will *always* be the best way to avoid them, but the rush of avionics technology has introduced two marvelous devices that are proliferating in the general aviation fleet. Based on significantly different principles, weather radar (a generic name for the kindred products of several manufacturers) and passive storm detectors deserve explanation, for serious instrument pilots will very likely be exposed to one or the other—or both—in the normal course of their experience. It's important for pilots to understand how these weather detectors work because of the inherent limitations and capabilities. First, the relatively new kids on the block: the passive storm detectors.

## Passive Storm Detectors

A number of years ago, a certain midwestern pilot who was transporting his family in their single-engine airplane managed to blunder into a thunderstorm that nearly did them in. Badly shaken, but determined to make something worthwhile out of the encounter, the pilot set out on a quest for an affordable, lightweight thunderstorm detector that would be compatible with light airplanes, and which would help other instrument pilots avoid severe weather. He just happened to be well educated in electronics, so he left his business, returned to school to study and develop his idea, and subsequently invented the Ryan Stormscope, the first of the passive storm detectors.

The principle of the Stormscope (and in general, other passive detection systems) is simple enough: It senses electrical discharges that are associated with atmospheric turbulence. Each discharge sensed by the Stormscope's antenna is processed electronically, and those that bear the distinctive signature of a turbulence producer show up as a bright green dot on a cockpit indicator. Range can be changed at the flick of a switch, with the result that you can observe a 360-degree, 200-mile display of probable turbulence around your aircraft.

Operation of a passive detector is as simple as its principle: You turn it on and take a look. In the air, weather avoidance is a matter of not flying into any area where the turbulence symbols appear. Unlike weather radar (which only shows areas of precipitation), passive detectors identify areas of probable turbulence, with or without the presence of rain. A strong "generator" (most likely a full-blown thunderstorm) may produce a phenomenon called "radial spread," in which an elongated string of symbols may grow from the storm location toward the center of the display, which represents your location. The appearance of radial spread should be a warning flag, prompting you to remain clear of that area.

Although some passive detectors provide full-circle viewing (with which you can see problem areas behind you, a big help when trying to decide which is the best way out of a stormy environment), your interest will most likely focus on what's ahead. To accommodate this, the controls may be adjustable to concentrate all the available symbols in the forward half of the display—a very beneficial feature in a weather-avoidance situation.

Because there is no outgoing energy to harm humans or to be affected by buildings, mountains, or the curvature of the earth, passive detectors are very useful for planning weather avoidance before you leave the ground. You can power them up on the ramp and get a good idea of turbulence to be expected, at least in the first 200 miles or so of your flight, and that helps eliminate surprises!

Experience based on frequent, regular observation of passive-detector indications will provide a data base for deciding how many dots, and in what concentration, are tolerable for you and your airplane. Learn to compare the cockpit display with actual conditions when you can see the weather, and you'll learn to evaluate and select safe weather-avoidance routes when you *can't* see the storms.

In summary, when detector #1 (the eyeball) is rendered unusable by clouds or the dark of night, the pilot using a passive detector who wishes to avoid severe weather need only refrain from flying where the turbulence symbols appear on the display.

## Weather Radar

Radar, in contrast, is hardly new; it's been around since early in World War II, when the German military began using it to detect Allied fighter aircraft swarming off Great Britain to intercept Nazi bombers headed across the English Channel. Radar is an acronym which stands for RAdio Detection And Ranging, and the modus operandi is relatively simple. Electrical energy in the form of radio waves is transmitted by a directional antenna, and some of that energy that strikes reflective objects is returned to the antenna, which is now functioning as an energy *receiver.* "Reflective objects" include the surface of the earth, buildings, mountains, and *precipitation*—that's the important reflector for weather avoidance. The return is displayed as a bright area on a cathode-ray tube in the cockpit.

While this isn't the place for a detailed electronic explanation, imagine the speed of radar's outgoing radio waves slowed to a crawl; when the antenna emits a burst of energy, the transmitter portion shuts down, turns the antenna into a receiver, and waits a predetermined time for an "echo" to return. If something reflects a portion of that energy, the returning radio wave is processed electronically and shows up on the radar scope, which is overlaid with a grid to indicate direction and distance from the aircraft. If *nothing* is reflected, the radar set switches back to the transmitter mode and repeats the cycle. In reality, all of this happens continuously and at a mind-boggling rate so that an uninterrupted display is provided. It's analogous to a light bulb powered by the 60-cycle alternating current in your home; the light is blinking on and off 60 times each second as the current reverses, but it's happening so rapidly that you perceive a steady glow.

A radar antenna sweeps back and forth—left-right, left-right— usually in a 90-degree arc with the airplane position at the apex. As echoes are "painted" (displayed on the scope), they remain illuminated through at least one antenna sweep so that you see a constant, updated display; the shape of an area of reflective objects is therefore easy to see.

Radar incorporates a third dimension of weather investigation; in addition to azimuth and several selections of range, the antenna can be tilted up or down as it sweeps and can "look" for reflective objects above and below the aircraft's flight level.

Atmospheric turbulence is generally associated with high rates of rainfall and large raindrops, both of which are characteristics of thunderstorms. Because water reflects radio energy very well, and because a weather system producing large amounts of rainfall can normally be expected to produce large amounts of turbulence as well, radar systems indicate the "severity" of echoes (in terms of potential airplane problems) in shades of green or by using several colors.

For example, if there are no reflective objects in the "field of vision" of your radar set, the screen will be blank; but let's generate a cumulonimbus cloud directly ahead, a cloud that is producing light rain. This precipitation would show up as a light green echo, its shape/azimuth/range clearly defined. Imagine this cloud growing to the point at which a core of heavier rainfall develops within the storm. When the reflectivity of the core rainfall increases to a preset level, the radar displays it in a brighter green, or perhaps yellow (for *caution*) in a multicolor system. There's usually a third level (even *brighter* green or red for *warning*) to indicate the highest preset reflectivity.

Some radar systems incorporate "contouring," which means that when the rate of rainfall increases rapidly over a relatively short horizontal distance the radar "closes its eyes" to that area. Should you be flying toward that storm mentioned earlier, and if the difference in rainfall intensity between the light rain and the core is increasing at a preset rate, the core would appear as a black hole (same as no return) on the radar scope. Pilots unaware of the significance of contouring might plunge directly into the heart of the thunderstorm, thinking they were headed for a no-rain, nonturbulent area. Surprise!

To call your attention to those weather systems that contain strong cores of precipitation and which therefore *must* be avoided, most radars cause contouring areas to flash on and off (in either green or red) as an additional warning. While some areas of precipitation displayed on a radar scope present no problem to most pilots in most airplanes, a "contouring" echo is all but guaranteed to rattle your teeth; you wouldn't like it one bit in there. The probable presence of very strong updrafts and downdrafts in close proximity to each other is the essence of life-threatening turbulence.

In the absence of very strong storms, radar will normally be capable of pointing out precipitation well in advance of the airplane because some of the electrical energy passes through to be reflected by another rain area up ahead. But can you imagine a big storm straight ahead that is producing so much rain that *all* of the radar's energy is reflected back to the antenna? You'd see a very bright echo to be sure, but what you *wouldn't* see is the big storm lurking on the other side. This characteristic is called "attenuation" and is a shortcoming of many older, less expensive units (some current-generation radars include circuitry that overcomes attenuation, or at least warns the observer of potential trouble behind the next echo). If nothing else, just knowing that attenuation exists should make it clear that you must have some idea of the extent and severity of a weather system before putting all your weather-avoidance trust in what you see on the scope.

There is a great deal of interpretation required of the pilot to extract useful, safe information from a weather-radar display. It's much more than a turn-it-on-and-take-a-look system because the picture changes remarkably with range selection, antenna tilt, airplane altitude and attitude, and storm characteristics. If radar is to be used with safety and efficiency, some well-founded training beyond reading the manual is required. There are frequent weather-radar classes and workshops around the country; your best source of information in this regard is the manufacturer.

Airborne weather radar and passive storm detectors are sometimes placed in a competitive posture, but any attempt at comparison soon turns into an apples-and-oranges standoff. These two weather detection systems operate on totally different principles and present totally different displays. However, improvements in technology and miniaturization have resulted in on-board weather-detection systems that combine the information from both types of detectors in one display, which is truly the best of both worlds. And if you'd like to spend even more money to stay out of turbulence and storms, you can buy weather-radar systems that use another color (usually magenta) to indicate the presence of turbulence as detected by the movement of reflective particles in the atmosphere. Still haven't flattened your wallet? You might look into one of the on-board radars that lets you push a button and look at a vertical slice of a storm; that's as close to a three-dimensional picture as you're likely to get.

Wise pilots *avoid* threatening weather (heavy precipitation and/or potential turbulence) that shows up on the equipment they have to work with. And the key word in that last sentence is *avoid*. They haven't yet built a weather-detection system good enough to pick your way through a squall line or to decide with confidence which of those storms up ahead is safe to penetrate. There's always another route or another day; if you become a faithful follower of the weather-*avoidance* religion, your chances of becoming an old pilot increase significantly.

# 12

# Communications Failure

It doesn't happen very often, but when it does, complete communications failure can really cause an instrument pilot to come unglued. If your radios are going to quit, they'll probably do it when you're fighting through a cold front, carrying an inch of ice, and wallowing around in moderate turbulence with a rough engine. This is no time to be digging through your flight bag, trying in vain to find the communications failure rules that you know are in there somewhere. You had promised yourself just last week you would review and really commit to memory those few simple regulations. And now it's happened, you don't know what to do, and there's always the chance that, during your confused wanderings about, you may crash into an innocent victim of your procrastination and lack of good sense. The point of all this should be unmistakable: Don't put it off, but learn (or maybe relearn) the rules that govern your actions when you lose communications.

Instrument pilots should toss a bouquet to the regulation writers at this point because the communications-failure procedures are clearly spelled out, with little room for misinterpretation. An obvious pattern of common sense is woven throughout this section of the rules; you will find that they *require* you to perform as you would instinctively, given the considerations of your safety and the well-being of other pilots sharing the airspace with you.

These rules are rather discriminating; they apply only to a failure of *communications*. When you can neither transmit nor receive on any radio, you have suffered a bona fide communications failure, and the rules spell out exactly what is to be done. If everything quits (no communications, no navigation) it's an entirely different ball game—sorry to break the news to you, but there are no rules for this situation. You're on your own; heaven help you, break out the common sense and good luck, and here's hoping you know which way to VFR conditions. (Always insist that the Weather-Depiction Chart be included in your briefing for an IFR flight; with that knowledge, at least you'll have an idea of which direction to fly.) Should ATC suspect that you have suffered radio failure, they will try to contact you on several possible communications channels; for example, the first place they will try to get through is the VOR that they figure you are using for navigation at the time. So if you suspect a radio problem, turn up the audio side of the VOR receiver, and listen; there may be enough power left for you to hear someone trying to help you. Of course, if you can hear, common sense tells you to listen and do what you're told.

Here's where a transponder takes on a new role. Suppose you have one enervated receiver weakly informing you of ATC efforts to make contact following loss of your transmitter. "Barnburner 1234 Alpha, if you read me, squawk *ident,*" which you do, and then hear, "Roger, Barnburner 1234 Alpha, radar contact, turn right heading one eight zero for radar vectors to a surveillance approach at Cedar Rapids Municipal Airport." And you would do just what the controller says; as long as the transponder and one receiver hold up, you've got it made—you're communicating.

There are other radio facilities on which ATC will attempt contact when your transmitter goes out, and they include Localizers, Outer Compass Locators, some NDBs, and any other navaid they think you might be able to receive. You should know that any navaid shown on an IFR chart (enroute, area, or approach) can carry voice signals unless the frequency is *underlined.* The essence of all this is that you should try anything you can think of when you suspect a radio failure: reset circuit breakers, check fuses, kick the panel, listen on any set available. You might even put the relief tube to your ear.

In a communications-loss situation, a hand-held transceiver changes from a gadget in your flight bag to a lifesaver. It's not as powerful as installed transmitters and its range is limited unless you can connect it to an external antenna, but a hand-held will likely enable you to communicate with someone who can help. If you can talk to any ATC facility or another pilot who can relay messages, you have reestablished communications and you can continue as if nothing happened. As a matter of fact, if you can do no more than *hear* questions and instructions from ATC, you can communicate with your transponder, and there should be no problem completing the flight.

If you are unable to establish any type of communications, don't panic; the rules will see you safely through this predicament.

# ✝ Out of the Clouds, Out of Trouble

The first rule was written in recognition of the undeniable fact that breaking out into VFR conditions will go a long way toward solving your communications-loss problem. If it happens in VFR, or if you fly into VFR conditions, it makes good sense to *stay* VFR, land as soon as practical, and let ATC know what you have done. Emphasis is on that phrase "land as soon as practical" because the terrain over which you are flying has a lot to do with your decision, and as always, safe operation must be placed before anything else. If the VFR into which you have flown is a hole in the clouds with Pike's Peak sticking up in the middle, you might do well to keep right on flying.

But when the radios quit in IFR conditions and it's down to minimums all the way home, you are on your own. What to do now? The answer is simple: Follow the rules.

When controllers suspect after several unanswered calls that you are experiencing radio problems (don't forget that you should announce your problem by squawking 7700 and then 7600 on your transponder), they are bound by the same set of rules to clear out some airspace ahead of you. It's obvious that the controllers need to know what route you'll be flying, how far you will go, and at what altitude. The only way to accomplish safe separation in the absence of communications is for both parties to operate under a set of common and inviolate conditions; that's why IFR pilots must know the radio failure rules inside out before venturing into the clouds.

It's easiest to break the problem into its basic parts. When communication is lost, you will have to decide:

1. How far you can go.
2. What route you should fly.
3. How high you should fly.

These three questions and their answers provide the guidance that will get you safely out of trouble; the problem is whittled down to manageable dimensions with these solid guidelines. One at a time, then.

# ✝ How Far Can You Go?

The first question has a most uncomplicated answer; unless you get lucky and fly into some VFR, you're going to go all the way to the destination airport for which you filed! If your clearance at the start of the trip (or a later amendment) specified the destination airport as the clearance limit, just keep right on going; you are still cleared all the way. But suppose you accepted a short-range clearance so you could get started, and now the radios have quit short of that point. No big problem, as long as you were provided an expect-further-clearance time. What has just happened to you is the only reason for these times being issued by ATC; in the event of a communications loss, the EFC is set far enough ahead to give the controllers time to clear out the airspace so you can continue on your way. This leads directly into the next phase of the procedure: You will remain at the clearance limit fix (holding in a standard pattern on the course on which you approached the fix, unless there is a holding pattern depicted on the chart) until the EFC time rolls around, and then you will proceed.

It should be clear by now that IFR pilots who accept a hold without getting an EFC leave themselves wide open for a real problem. If the controller doesn't issue an EFC time, *demand* one; you're well within your rights. Picture the bewilderment on the ground and in the cockpit when the radios go out and no further clearance time has been provided; they don't know that you don't know that they don't know, and it's a scramble to get everybody out of the way on all the possible routes you might take at any time.

So you were cleared all the way to destination, with no clearance limits, and the radios gave up the ghost soon after takeoff? In this situation, proceed to your destination and hold at the appropriate altitude until your estimated time of arrival, based on the time enroute you put in your flight plan and your takeoff time—which, of course, you have forgotten. During a period of mental stress such as you may be experiencing following a radio failure, it may become very difficult to remember your own name, let alone what time you took off! Here's where you will be forever grateful for having formed a most useful habit; just before takeoff, set the red hands on your instrument panel clock, or on that 3-inch diameter, $1000 wristwatch that does everything but compute your income tax, or write it down somewhere, but record the takeoff time. Obviously, it is also nice to remember the time enroute you filed in your flight plan.

# ☩ What Route Should You Fly?

The first part of the communications rule directs you to continue to your destination, respecting any holds or clearance limits along the way; now you are faced with the question of what route to fly. Equally sensible, this part of the problem is solved in one of three possible ways: You will fly via the last assigned route, the route ATC advised you might expect in a further clearance, or in the absence of the preceding two, the route you filed in your flight plan.

To illustrate, suppose you plan a trip from Lexington, Kentucky, to Bluefield, West Virginia (Fig. 12-1). You requested Victor 178 to

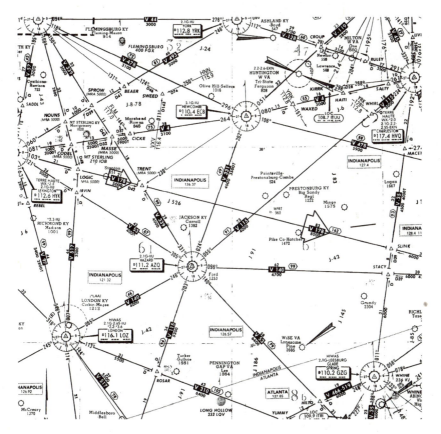

**Fig. 12-1.** *Airways and navigational facilities between Lexington, Kentucky, and Bluefield, West Virginia. (Copyright Jeppesen Sanderson Inc., 1979, 1997; all rights reserved; not to be used for navigation)*

Bluefield, and your clearance comes through "as filed." If radio failure occurs anywhere along the route, you will continue on Victor 178 to the Bluefield VOR. So much for the first situation.

In the second case (same route, same clearance), you have just passed LOGIC intersection when Indianapolis Center comes up with this little gem: "Barnburner 1234 Alpha, you are now cleared to the TRENT intersection, maintain 7000 and expect further clearance at 46 via Victor 339 Hazard; Victor 140 Bluefield." A couple of minutes after acknowledging this clearance, your radios give up. Combining your knowledge of the first two communications-failure rules, you will hold at TRENT until 46 minutes past the hour and then proceed to Bluefield via Victor 339 Hazard and Victor 140; that's the clearance ATC advised you to expect, so it takes precedence.

Back up one more time and copy the following clearance on the ground at Lexington: "Barnburner 1234 Alpha is cleared to the TRENT intersection via radar vectors to Victor 178, climb to and maintain 7000, expect further clearance at 28." The implication is that your flight cannot be cleared all the way to Bluefield at this time, but ATC is willing to get you started with a clearance as far as TRENT intersection, with the rest of the routing to follow. You accept, and as soon as Departure Control has vectored you onto Victor 178 10 miles from LOGIC, the radios die. Proceed to TRENT, hold until 28, and then continue down Victor 178. In the absence of an assigned routing, without even a routing to *expect*, you're expected to fly just as you requested when you filed your flight plan.

ATC will frequently direct you around other traffic, or thunderstorms, or if they're not particularly busy, they will sometimes vector you all the way home. Should radio failure rear its ugly head at this point in an IFR flight, you will proceed directly to the fix to which you were being vectored (using your own navigational skills, of course). Good judgment once again prevails, as it would be rather stupid to turn and fly into a thunderstorm in order to go direct to the next VOR, according to the rule. If you were advised of a storm cell 10 miles wide straight ahead and then were given vectors around it, common sense would dictate continuing the vector until you are reasonably certain you are past the storm. Don't ever let *anything* or *anybody* force you to fly through a thunderstorm. And don't worry about traffic separation if you wisely elect to continue around the storm after the radios stop radioing; you had to be in radar contact to get the vector in the first place, and the controller would probably breathe a sigh of relief, watching you exhibit some "aeronautical smarts."

In summary, you are expected to fly one of these three routes when the thread of communication snaps (they are listed in order of precedence):

1. The route you have been *assigned* as part of a clearance.
2. The route that ATC advised you might *expect* in a further clearance.
3. As a last resort, the route you *filed* in your flight plan.

(When radar vectored, proceed directly—or as appropriate—to the fix to which you were being vectored.)

## ☨ How High?

Only one more question needs an answer to solve this three-dimensional problem of what to do when you realize you're talking to yourself, and that's the one regarding altitude. Your safety is the big concern here, so the rules require you to fly at the highest, and therefore the safest, of these altitudes:

1. The last assigned altitude.
2. The altitude ATC has advised you may expect in a further clearance.
3. The Minimum Enroute Altitude (MEA) for the particular airway segment in which you are operating.

To illustrate, consider the same Lexington-Bluefield trip; you were cleared "to the Bluefield Airport via Victor 178, climb to and maintain 9000." If radio failure happens anywhere along this route, you will remain at 9000 until reaching Bluefield (notice that there are no higher MEAs on this route), hold until your ETA, and make the approach. You are obliged to stay at 9000 until your time runs out and then to descend in the holding pattern prior to executing the approach. (By the way, in a situation like this, the approach you use is entirely up to you; anything published is legal.)

Had you requested 9000 but ATC was unable to grant it in time for your takeoff, they might have cleared you "to the TRENT intersection, maintain 5000, expect 9000 after TRENT, expect further clearance at 36." You take off, climb to 5000, and the radios quit before you reach TRENT intersection. Hold there until 36 minutes past the hour, depart TRENT climbing to 9000, and maintain that altitude all the way to Bluefield. Again, you're above all the MEAs on the airway, and since you had been advised to expect 9000 feet after TRENT, that's the altitude you fly.

But maybe the airways are so congested that Center can clear you only to LOGIC intersection at 5000 with no promise of the necessary higher altitude at that time. Your task is to determine if there are any MEAs *higher* than your last assigned altitude of 5000. Sure enough, you'll need to climb to 8000 feet for the segment between TRENT and SLINK and then descend to 6000 for the final segment into Bluefield.

You are expected to fly altitudes *as published* on the enroute chart when they apply. For example, in the absence of an assigned altitude, you would fly at 4700 feet from Hazard to the STACY intersection. East-odd, west-even, and the "nearest thousand" don't count here—fly 'em like you see 'em!

Summing it up, your determination of what altitude to fly is rather simple; it's the last assigned altitude or the MEA, *whichever is higher.* If ATC has advised you to expect a higher altitude enroute, that's the one that applies. Whenever a route segment calls for an MEA higher than anything you have been assigned or advised to expect, start climbing to the higher MEA over the fix that begins that segment and begin descending to the previous highest altitude over the fix that terminates the segment. Airway segments are defined by enroute VORs and elsewhere by short perpendicular lines—T-bars, if you will—on the airway centerline. When you see a VOR or a T-bar on the airway, look for a change in MEA. (When it's unsafe because of terrain or obstructions to start climbing at the beginning of an airway segment, the intersection will be clearly marked with a *Minimum Crossing Altitude* symbol, requiring you to be at or above a specified level before proceeding. You should climb in a racetrack holding pattern or start climbing before reaching the intersection.)

When dealing with a communications-failure situation, there's one more area in which pilots and controllers must agree; you must not comply with the conditions of a clearance unless you are reasonably certain that ATC has received your acknowledgment. If you have any doubt, proceed as if you had never heard the instructions because a clearance is considered valid only when it is acknowledged.

Here are the regulations in capsule form. When you lose two-way radio communications while in IFR conditions, you will:

1. Maintain VFR and land as soon as practical if the failure occurs in VFR conditions or if VFR is encountered after the failure. (This rule is to be salted heavily with common sense and good judgment.)

2. Route: Continue to your destination on the last assigned route *or* a route ATC has advised you may expect *or* the route filed in your flight plan. When being radar vectored, proceed directly to the fix.

3. Distance: Proceed to the clearance limit and hold at the appropriate fix until the EFC or EAC time; then continue on your route. If the clearance limit is the destination airport, hold at the last assigned or appropriate altitude until your ETA, based on takeoff time plus ETE.

4. Altitude: Maintain the last assigned altitude or the MEA, whichever is higher, or at the prescribed fix, climb to the altitude ATC has advised you to expect. If subsequent MEAs are higher than the assigned or expected altitude, climb to the higher MEA for the particular airway segment and then descend to the former highest altitude.

These rules work. Don't expect to be treated like a hero at the completion of a no-radio IFR flight because the controllers know what you will do and make traffic adjustments as required. But this mutual assistance pact works only if pilots and controllers follow the rules. So hold up your end of the bargain: Take time now and at frequent intervals to memorize and refresh your understanding of the communications-failure procedures; when it happens, it's too late to learn.

# 13

# Holding Patterns

"This is Mercury Control; we are at 'T' minus 30 minutes and holding." That was the voice of Colonel Shorty Powers, announcing a hitch in one of the early orbital flight schedules and enriching the public vocabulary with a Space Age connotation of the word "hold." It was about time the space program caught up because holding has been with us in instrument flying for years as an orderly way of suspending the progress of an instrument flight, either enroute or in a terminal area. Since most aircraft depend on forward movement to keep from falling out of the sky, holding really becomes a matter of flying around in circles without going past a predetermined point. The reasons for holding are numerous and sometimes apparently known only to heaven and ATC. But it provides an effective, if frustrating, way to "stop" the flow of IFR flights in the face of a traffic conflict or to slow things down when terminal operations begin to exceed the capabilities of the system.

Being cleared to hold sooner or later is just as inevitable as automatic rough over the mountains at night, and with Big Brother watching on radar, you'd best know what you're about! In B. R. (Before Radar) days, you could enter and fly a holding pattern any way you pleased, as long as you stayed within the airspace limits; but on the very first day the radar sets were plugged in, controllers observed a succession of pilots entering holding patterns and were overwhelmed by the serpentine tracings of their entry procedures. So overwhelmed, in fact, that they were moved to call them "small intestine entries." This led to the development of an official method of holding pattern entry.

There are basically two situations in which ATC will require a flight to hold: one during the enroute portion of a trip and the other in the terminal area, in which case you will be holding in anticipation of an instrument approach to the airport. The rules are the same for both; they differ only in the procedures you use in the event of communications failure.

No matter where or when you are instructed to hold, there is a two-step method you should use to do the job properly. Step #1, decide where the "racetrack" pattern will be located in accordance with the instructions you have received (you can do this mentally or actually draw it on the chart). Step #2, based on the direction from which you are approaching the holding fix, determine the entry procedure you will use.

Every holding clearance must include either specifically or by implication these four components: a *holding fix,* the *direction to hold,* the *type of pattern* (standard or nonstandard), and an *expect-further-clearance* time. Each part of the holding instructions should be considered by itself.

## ┼ The Holding Fix

ATC can clear you to hold just about anyplace they want to, but the holding fix will always be one you can locate electronically; it could be a VOR, an NDB, an intersection, a DME fix, or an RNAV waypoint. Most enroute holds will be at intersections or VORs along the airway, and terminal holding fixes are usually set up so that you will be in position to commence the approach when it's your turn.

## ┼ Direction to Hold

Before attempting to unravel the mysteries of "which way to hold," you must understand what is meant by the holding course. For example, the Big Piney VOR has been designated the holding fix (Fig. 13-1), and you are cleared to hold east of the VOR (in the absence of more specific instructions, this means you are to hold on the 090 radial). Fundamental to solving the problem is the knowledge that the *holding course* lies east of the fix. Remember that the holding course always lies in the direction you have been cleared to hold, and once in the pattern, you will fly inbound to the *holding fix* on the *holding course* every time you come around the racetrack. True, you'll be flying westbound when on the holding course, but remember that "holding east" is just a means of identifying the location of the holding pattern.

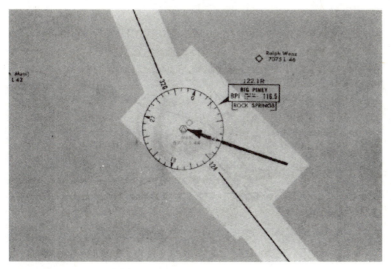

**Fig. 13-1.** *The holding course always lies in the direction you were cleared to hold.*

# ✝ Type of Pattern

With the holding fix, the holding course, and the direction to hold firmly established, determine the type of holding pattern to be flown—either standard or nonstandard. For the time being, consider right turns standard and left turns nonstandard. If the controller doesn't mention the direction of turn, you're expected to execute a standard pattern with right-hand turns. To complete step #1, just "fly" your pencil inbound to the holding fix on the holding course and turn in the appropriate direction. For example, to hold southwest of the Big Piney VOR in a standard pattern (Fig. 13-2), find the holding fix and establish the holding course; then "fly" inbound to the VOR on the holding course and turn right. Proceed 1 minute outbound, turn right again, and when you arrive over the VOR, you've completed one circuit of this particular holding pattern. If the controller clears you for left-hand turns, fly inbound on the same holding course and turn left; you're still *holding* southwest of Big Piney.

# ✝ Which Entry to Use?

With the holding pattern established, proceed to step #2, determining the entry procedure. Fortunately, the one used most often is the easiest to accomplish. Suppose you are flying northwest at 10,000 feet on

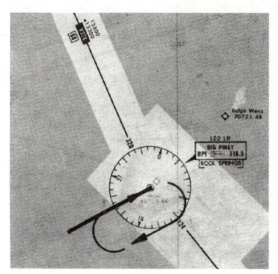

**Fig. 13-2.** *Standard holding pattern (right turns) southwest of Big Piney VOR.*

Victor 328 and are given this clearance: "Barnburner 1234 Alpha, hold southeast of Big Piney on Victor 328, maintain one zero thousand, expect further clearance at four two." If you want to copy the important numbers, do it right on the chart (Fig. 13-3), and you'll have a permanent record of the clearance. Next, break the clearance into its component parts:

1. The holding *fix* is the Big Piney VOR.

2. The holding *course* is Victor 328.

3. Since you will hold southeast of Big Piney on Victor 328, the holding course will lie southeast of the VOR on the 124 radial.

4. Absence of instructions to turn left or right implies a standard pattern (right-hand turns).

5. You will continue to fly at 10,000 feet while holding.

6. If you should suffer complete loss of communications before receiving further instructions, you will be expected to resume your flight at 42 minutes after the next hour.

Now go ahead with step #1, establishing the pattern (Fig. 13-4). Draw (or imagine) the holding course southeast of Big Piney on Victor 328, fly your pencil inbound to the fix on the holding course, turn right to 124 degrees, and you have the proper racetrack all set up; you're holding southeast of Big Piney on Victor 328.

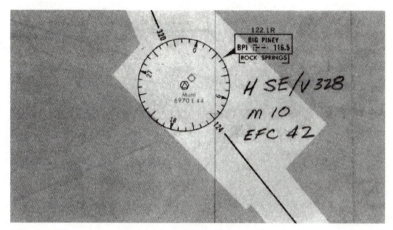

**Fig. 13-3.** *Copy your holding clearance right on the enroute chart.*

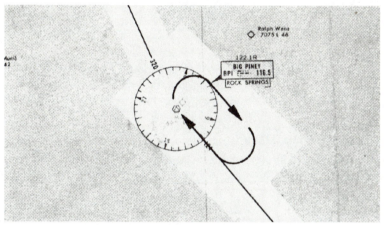

**Fig. 13-4.** *Holding in a standard holding pattern southeast of Big Piney VOR on Victor 328.*

## The Direct Entry

When you fly to the fix and turn to the outbound heading, you have accomplished a "direct" entry, which is always used when you're on the holding course as you approach the holding fix. As a matter of fact, your approach heading can be somewhat off the holding course and the direct entry will still be the proper one to use. If you draw (or imagine) a circle around the holding fix and then draw a line that lies 70 degrees from the holding course (Fig. 13-5), a direct entry is called for whenever your heading puts you in the shaded half-circle. As soon

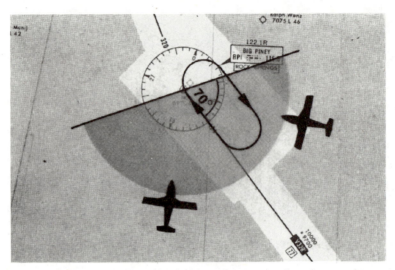

**Fig. 13-5.** *Approaching the holding fix on any heading within the shaded half-circle calls for a direct entry.*

as you pass the VOR, turn right to 124 degrees, and you have entered the holding pattern. The 70-degree line is the key to determining your entry procedure. It must always be drawn (or imagined) so that it cuts across the racetrack, which you established in step #1.

## The Teardrop Entry

Keep the same holding instructions in mind (hold southeast of Big Piney VOR on Victor 328), but this time you are approaching from the west (Fig. 13-6). Set up the racetrack again, draw the 70-degree line, and extend it through the fix. While you're at it, extend the holding course through the fix as well, which divides the unshaded half of the circle into two quadrants. The smaller of the two is of concern now, since you are approaching the fix on a heading that lies within that quadrant. When you reach the fix, a right turn to the outbound heading (124 degrees) would put you very close to the holding course but going in the wrong direction, so a procedure has been designed to get you into the holding pattern quickly and well within the limits of the airspace set aside for you. In this entry, turn to a heading no more than 30 degrees from the holding course (which in this case would be a heading of 94 degrees), fly for 1 minute, turn right again until you intercept the holding course, and return to the fix. (Don't forget to reset the OBS to 304.) This is the "teardrop" entry, the name coming from the flight path described.

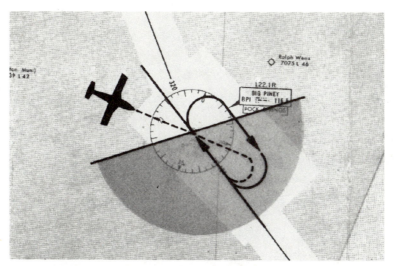

**Fig. 13-6.** *The teardrop entry is used whenever you are inbound in the smallest sector; turn no more than 30 degrees from the holding course.*

## The Parallel Entry

The third entry—the "parallel" entry—is used when approaching the fix on a heading which lies in the larger of the two unshaded quadrants. Suppose you have received the same holding instructions, but you're approaching Big Piney from the north (Fig. 13-7). When you reach the fix, turn to the heading which *parallels* the holding course (in this example, a heading of 124 degrees), fly for 1 minute, and then turn left until intercepting the holding course or until you reach the fix, whichever happens first. A stiff tailwind as you turn inbound will often push you direct to the fix before you intercept the holding course.

# ✈ Making It Easier

The procedures described thus far are the ones officially recommended by the FAA, and they provide a near-absolute guarantee that you will remain within holding airspace during the entry to a holding pattern. The 70-degree line and the direction of the turns were developed with high airspeeds and adverse wind in mind, and were carefully calculated to take care of even the worst set of conditions. Pilots of high-performance aircraft—turboprops and turbojets operating at high altitudes where strong winds can blow them off course in short order—have legitimate

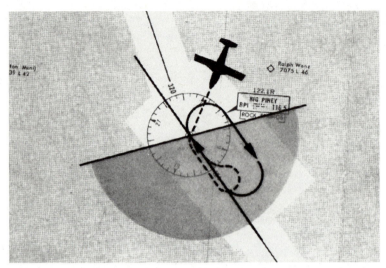

**Fig. 13-7.** *The parallel entry works when neither teardrop or direct entry procedures apply. At station passage, turn to the reciprocal of the holding course.*

concerns about drifting outside protected airspace. But for lightplanes at lower altitudes, displacement due to wind and radius of turn doesn't count for much, and a simplified holding pattern entry can be used.

To make it easier, when you arrive at the holding fix, *turn in the shorter direction to the heading that parallels the holding course.* Fly outbound for not more than 1 minute, proceed direct to the holding fix, and start around the racetrack in the appropriate direction. (When it's obvious that such a turn would place you on the side of the holding course away from the racetrack, don't do it; turn the other way. For example, the westbound aircraft in Fig. 13-6 should turn to the *right* when reaching Big Piney VOR even though it's the long way around. A left turn to 124 degrees would put the airplane on the nonholding side, where the protected airspace is considerably smaller; you're not supposed to be there while holding.)

As soon as you're squared away on the outbound leg, rotate the OBS to match the outbound course, and the CDI will indicate whether you should turn left or right to get to the holding course. With an HSI, put the course pointer on the *inbound* number, since the movable course segment is always directional.

During the first minute outbound, adjust your parallel heading for wind drift. A little "Kentucky windage" goes a long way here—5, 10, 15 degrees into the wind, based on what you experienced while tracking toward the fix. And when you turn back toward the fix at the end of the minute, be aware of your relation to the holding course; it's quite likely that you will fly right onto it during the turn.

Even though there's no regulatory basis for it, some instructors and examiners insist on the use of the FAA's entry procedures (they are *recommended,* not mandated). Prior to a flight check, pilots who prefer the simpler entry described here (or one of their own design) should discuss their procedure with the examiner so that there's no misunderstanding. However it's accomplished, a holding pattern entry should keep the airplane close to the holding course, never more than 1 minute or the specified distance away from the holding fix.

# ⊥ Holding Pattern Limits

When you are cleared to hold, the controller assigns you a block of airspace large enough to separate you from other IFR traffic during your entry and subsequent pattern maneuvers. Airspeed limits are placed in the regulations to guarantee that aircraft entering a holding pattern will not go swooping outside the boundaries set up for them. For propeller-driven planes, the maximum speed is 175 knots indicated, and you must slow to this speed (or less) within 3 minutes prior to crossing the holding fix. Even with a strong tailwind, 175 knots will see you safely through the entry and the holding pattern.

If your airplane is powered by turbojets (that is, no propellors), you may increase the maximum speed: 200 knots up to 6000 feet, 210 knots between 6000 and 14,000 feet, and 230 knots above 14,000 feet (all indicated airspeeds, of course). In any event, be at or below the maximum holding speed before arrival at the fix.

There is one more variable in this airspace guarantee, and that is the angle of bank. If all aircraft observe the maximum holding airspeeds, the lateral displacement during turns will depend on how rapidly those turns are accomplished. Therefore, during the entry and while holding, accomplish your turns at the standard rate—3 degrees per second—unless that rate would require an abnormal angle of bank. It's easy to figure out why you might want to cut down the rate of turn a bit when you consider the direct relationship between true airspeed and the angle of bank needed to generate a standard-rate turn. You can approximate the bank angle by this rule of thumb: Divide the true airspeed (in knots) by 10, add 5, and you're pretty close to the necessary bank. (Applying this to an actual situation, a jetliner flying at 600 knots TAS would have to be racked into a 65-degree bank to turn at standard rate. Although the airplanes can take it, people would complain—American industry builds 2-g airplanes, but not 2-g passengers.) When high true airspeed is involved, bank not more than 30 degrees or whatever's normal for your autopilot/flight director.

# ✝ It's "Four T" Time Again

Approaching Big Piney VOR the CDI will begin to dance a bit, start its reversal, and when the TO-FROM indicator shows FROM, two things occur simultaneously; you have arrived at the holding fix and, for reporting purposes, you are in the holding pattern. Note the *time* as you pass the holding fix, *turn* to the outbound heading, *throttle* back to holding airspeed, and after all this is done, *talk*; you owe ATC a report of the time and altitude entering the holding pattern. There's no need to rush this report to the controller because you're doing exactly what's expected of you, but when you are established on the outbound heading and everything is under control, report: "Salt Lake Center, Barnburner 1234 Alpha, Big Piney at three two [the time you first crossed the fix], holding at one zero thousand."

Slow down in the holding pattern because you're going to be flying around in circles for a while, and it doesn't make good sense to burn up more fuel than necessary. Next time you're under the hood, try several power combinations and determine a comfortable holding airspeed for your airplane; make it a tradeoff between ease of handling and economy.

In the time it has taken to read this, you will have turned outbound and accomplished the four Ts, so in the piece of a minute you have left before turning inbound, rotate the OBS to the holding course (304 degrees) and you are ready to turn inbound to the holding fix.

Executing standard-rate turns during the teardrop and direct entry maneuvers in a no-wind situation should put you on the holding course when you roll out of the turn toward the station. The parallel entry sets you up for a 45-degree "cut" on the inbound course, but the gods who rule the atmosphere have seldom been known to bless pilots in holding patterns with calm winds, and the disorientation caused by wind drift can be significant. A clue to your position and the wind factor you're fighting shows up during the turn inbound, when your heading is 45 degrees from the holding course (in the example, 259 degrees). If at this point the CDI has not begun to move from its full left position, stop the turn and hold 259 degrees until the needle comes alive (see Table 13-1).

In this case, there is obviously a westerly component to the wind, and it has drifted you east while flying the outbound leg. Should the

**Table 13-1.** Course Intercept Situations and Solutions

| Situation* | Solution |
|---|---|
| CDI has not moved from full left or full right (depends on direction of turn) | Roll out, fly this "45-degree" heading until CDI centers |
| CDI starting to move toward center | Continue turn; you should roll out on course |
| CDI already centered, or well on its way | Increase bank to 30 degrees; prepare to apply correction to get back on course |

*At the 45-degree point.

CDI start sliding toward center while you're turning through 259 degrees, keep turning and you'll roll out on or very close to the course. Movement of the CDI *earlier* than the 45-degree point in the turn is a signal to increase the rate of turn (30 degrees of bank is plenty) and prepare to correct for an easterly wind.

These three situations and their solutions at the 45-degree point of the inbound turn will work only if the OBS is set on the holding course, which will cause the CDI to always be displaced toward the holding course.

# ✛ Timing the Holding Pattern

You are now on the threshold of the most important phase of holding pattern timing. The rules call for an inbound leg length (on the holding course) of 1 minute at or below 14,000 feet, and 1.5 minutes above 14,000. You are given complete freedom to adjust your timing on the outbound leg so that the inbound requirement is satisfied. The moment the CDI centers on the inbound course (or when you complete the turn), start your timer and note carefully how long it takes to return to the holding fix. If it turns out to be exactly 1 minute, you've done it; you have hit that rare combination of perfectly executed turns and absolutely calm winds.

But suppose a minute and a half slips by before you arrive back over the VOR. You have encountered some tailwind as you flew outbound, and it's cut-and-try from here on. When you start outbound the next time, try 30 seconds and then turn back to the holding fix and check the inbound time; it should come out very close to

1 minute. There will be times when the outbound leg will be only 10, maybe 20, seconds long; when the wind is *really* strong, you may find yourself doing what appears to be a 360 over the fix to keep the inbound leg at the proper length. On the other hand, you might have to fly outbound for 2 or 3 minutes, depending on the wind direction and force.

Once established in a holding pattern, your only concern with respect to timing is to make certain you're on the holding course *inbound* for 1 minute (or 90 seconds, depending on altitude)—no more, no less. Therefore, you must know where to punch the stopwatch as you turn outbound. The solution is simple: Start timing when you are abeam (directly opposite) the holding fix.

## ✛ Where Is Abeam?

There are several holding situations that require different methods of determining when you are in the abeam position. The easiest is holding on a VOR, where the timing point is the TO-FROM reversal (Fig. 13-8). Next in order of simplicity is a "square" intersection (VOR radials or a radial/ADF bearing combination crossing at or near a 90-degree angle), in which case you start the timer upon centering the side radial or attaining the proper relative bearing. When the references are not close to "square," you must begin timing at the completion of the turn to the outbound heading. This is also the best procedure when holding on an NDB; start timing when the outbound turn is completed.

It will likely require several circuits to nail down the exact timing, so continue experimenting until you arrive at an outbound leg length which results in exactly 1 minute on the holding course.

## ✛ Drift Correction

A technique called "double drift" is an effective way to compensate for wind in a holding pattern. When you have found a heading that will keep you on the holding course, correct *twice* that amount into the wind as you fly outbound. Your pattern will lose its symmetry (Fig. 13-9), but you will intercept the holding course handily and you'll stay safely within the airspace reserved for you. Combine good drift correction with proper timing and your holding patterns will be just what the doctor (ATC) ordered.

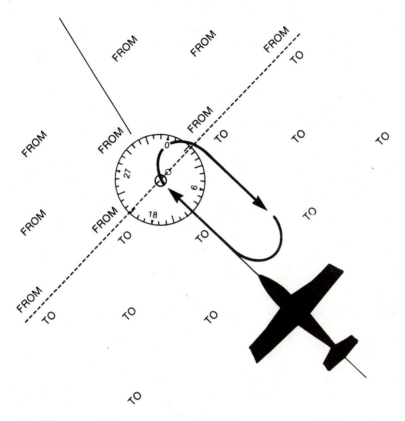

**Fig. 13-8.** *With the OBS set properly, the TO-FROM indicator provides a good timing device in the holding pattern.*

# ✝ How Long to Hold?

The final portion of any holding clearance will consist of the time at which you can expect further clearance. Don't *ever* accept a hold unless you receive an EFC; if a controller forgets, ask—nay, *demand* it! You will seldom have to hold until this time, since it is purposely set far enough ahead to provide a clear route for you if your radios quit (see Chap. 12, Communications Failure, for details). However, you should plan to arrive over the holding fix (or at least be inbound to the fix) when the EFC time comes up on the clock. It is perfectly legal—and often necessary—to shorten the racetrack to do this. Bear in mind that a no-wind holding pattern (at or below 14,000 feet) consumes 4 minutes per circuit, so it is a matter of shortening the outbound leg to accomplish station passage at the proper time. Make a complete turn over the fix if your EFC is only 2 minutes away

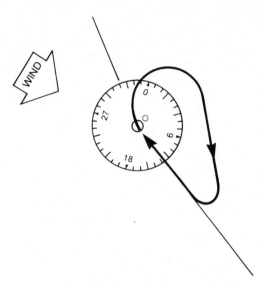

**Fig. 13-9.** *This is the track described by a holding aircraft using double drift to correct for the wind.*

when you cross the fix and be ready to go about your business on time.

Use the sometimes interminable minutes in an extended hold to listen to what's happening ahead of you. By paying attention to instructions and clearances issued to others in the holding pattern, you can get a good idea of whether you'll be cleared on your way or be held still longer. If your EFC is close (5 minutes) and no word is forthcoming from ATC, give them a call and find out what to expect; you have a right to know because continued holding may begin to impinge on your fuel reserve. You may want to divert to some less crowded route or terminal. *Any* hold should cause you to take a critical look at your fuel supply; remember that *you,* not the controller, are responsible for not running out of fuel.

An extended holding situation is a wearying exercise in making circles and going expensively nowhere, but sometimes it's unavoidable. When it appears that you're going to hold for a long while and you tire of going around every 4 minutes, you might consider requesting a nonstandard, longer-legged pattern of your own. This could be done by asking for 2- or 3-minute legs or by using a convenient DME fix to stretch things out a bit. If it can be done safely, ATC will usually go along with your request.

# ✝ Holding with DME

"1234 Alpha is cleared to hold northeast of the Jack's Creek VOR, 8-mile legs, maintain one three thousand, expect further clearance at one seven." This is the easiest holding pattern of all because a DME reading replaces timing. In this case, fly to the VOR, enter the pattern, and fly outbound until the DME reads 8; then return to the holding course, fly back to the VOR, and do it all over again. It may require a bit more attention to drift correction because you're not basing your pattern on time and the "double-drift" method doesn't always work, but the end of the outbound leg is very clearly indicated (8 miles northeast of the VOR), and you won't have to worry about timing.

# ✝ Depicted Holding Patterns

On occasion, the nice folks who figure out the whys and wherefores of instrument enroute charts eliminate step #1 of the holding pattern procedure by depicting the racetrack for you. Published patterns will be found at intersections frequently used for holding. They ease the workload of controllers and pilots because detailed holding pattern descriptions are unnecessary. If you need to be held at an intersection that shows a published pattern, the controller has only to say, "Hold at the so-and-so intersection," which is your cue to fly the pattern just as it's shown on the chart. All published patterns are assumed to be standard as far as timing is concerned; if the controller wants you to fly longer legs, you'll be advised.

# ✝ It Happens to Everybody

Drawing a holding pattern and the proper entry on a piece of paper and actually flying it are two different things, to which every honest IFR pilot will bear witness! The first few times will likely leave you wondering where the heck you are, and fortunately, there's an easy way out of this situation. When disorientation strikes, level the wings (no matter what direction you're headed), rotate the OBS until TO appears, and then turn to that heading and fly to the station. On the way, orient yourself to the radial you're on and when FROM shows up start all over again. Should you have to do this several times in succession, don't worry—at least you're staying close to the holding fix! If you become utterly and genuinely lost, holler for help, and ATC will come to the rescue. They're probably more anxious than you are to straighten things out.

# 14

# Getting Ready for an Instrument Approach

There are several kinds of instrument approaches in use today with more just over the horizon, but before beginning *any* approach, there are several things you can do to make the entire operation easier and safer.

In the interest of constantly being one step ahead of the airplane, be aware of events that usually signal the beginning of the approach phase. For example, the Center controller may clear you to a lower altitude to fit the requirements of terminal area traffic flow; you may be given the first in a series of radar vectors to put you in line for the approach; in a nonradar environment, Center will frequently hand you off to a controller at the airport, and you may be requested to report passing an intersection or a DME fix as you fly inbound. When you reach the specified point, you will probably be cleared for the approach. Listen to other conversations on the Approach Control frequency and you'll get a good idea of *what* to expect and *when* to expect it.

Find out as early as you can what approach is in use at the destination, especially if it's an airport or an approach that is strange to you. If you're bound for an airfield with only one published procedure, there's no problem, but the larger terminals may have several types of approaches to three or four different runways. At one time, Chicago's O'Hare International had 24 procedures, plus radar approaches to 8 runways.

There are several sources of approach information; at the top of the list is Automatic Terminal Information Service (ATIS), broadcast continuously on a VHF frequency listed on the approach chart.

When you're headed for a non-ATIS terminal, Approach Control will usually spill the beans something like this: "Barnburner 1234 Alpha, you're in radar contact; expect vectors to the final approach course for the Runway 16 VOR approach, last reported weather is…"

A Center controller can always get the approach and weather information for you, but you shouldn't make that request unless you need to know a long way out. ATIS or Approach Control notification should give you plenty of time to get set up for the approach.

What about the airport that has no ATIS and no Approach Control? If you don't want to bother Center, do the logical thing and monitor the Tower frequency for a while on your other radio; you'll likely find out what procedure they're using. Or ask Center if you may leave the frequency for a moment (a request that's almost always granted) and call the Tower, from whom you can also get the weather situation first-hand. If there is not even a Tower at Boondocks Municipal Airport, there's likely an AWOS facility or a Flight Service Station that can supply you with nearby weather and winds, and you can make up your own mind about how to get to the runway.

As soon as you are in the terminal area, put away your enroute charts, notepads, computers, coffee jugs, and whatever else might divert your attention from the approach chart. Clean up the cockpit so that you can devote all your talents to flying the approach.

# ✛ Good Practices Make Good Habits

As you study the published procedure, there is a method you should use for maximum efficiency, standardization, and safety. Done exactly the same way every time, it will soon become a habit and will prevent missing some vital bit of information on that dark, stormy night when the chips are down. In a stressful situation, your well-practiced method of chart study will come booming through.

Here's the way to do it: The first two items (and at this point the *only* two) to be memorized are the *first heading* and the *first altitude* of the missed approach procedure. Allow those two numbers to burn themselves into your mind because the next item on the approach agenda is to convince yourself that you will definitely have to execute a missed approach, even though you *know* that the airport is VFR. With this attitude, no major shift of mental gears is required on those very rare occasions when you are forced to go around because of weather. And what a pleasant surprise when you break out of the clouds and there's the runway, straight ahead; it may be the wrong one, but in bad weather *any* runway is beautiful.

That leads directly to the next part of your approach preparation: What will the airport look like when you make visual contact? That should be a no-sweat situation at your home field, but when you're executing an approach to a strange airport in weather all the way to minimums or when you're cleared for a circling approach to a runway other than the one with which you're lined up, the potential for disorientation increases significantly. Prevent embarrassment ("Tower, this is 34 Alpha, if you have me in sight, please tell me which way to turn") by studying the airport plan view on your approach chart. Turn the chart so that you're looking at the airport layout just the way it will appear when you break out of the clouds; find the landing runway and decide which way you will turn. It's also an excellent way to make yourself aware of trees, power lines, grain elevators, high-rise apartments, and other obstructions that people insist on building in approach airspace.

Next item for the study method is the approach itself. All published procedures have one thing in common; they proceed step by step from the initial approach through various segments to the landing or missed approach, as appropriate. At this time, make a "big picture" study of the entire procedure: headings, altitudes, distances, times, and so forth. When you actually start the approach, always let your mind fly one segment ahead of the airplane.

Now tune, identify, and orient yourself to the navigational aids you will use. You should set up your nav radios as soon as you get the first radar vector from the approach controller. Your only navigational responsibility at this point is to comply with the vectors. If the missed approach is an involved one, you would do well to set up your radio equipment so that there will be a minimum of knob-twisting when you're busy with a go-around.

After all this, you have one more number to memorize, and it's really an important one: the MDA or DH, depending on the type of approach in use. In either case, it's the lowest altitude to which you may descend until the runway is visible. Picture in your mind's eye how the altimeter hands will appear when you get to that altitude and keep it in mind. If you have the luxury of another aviator in the right seat, you can have that pilot let you know when you're 100 feet above the minimum altitude.

Approach charts are published on single pages and kept in loose-leaf binders for good reason; they are intended to be removed and placed somewhere in the cockpit for ready reference during an approach. The best possible location is in the jaws of a spring clip firmly attached to the control wheel so you won't have to move your head to look at it. (Next time you are under the hood, put your approach plate on the floor between the seats; just as you begin your turn back onto the final approach course after completing a procedure turn, lower your head and look down at the chart for 10 seconds; then go rapidly back to the panel. This will probably induce vertigo such as you never believed could happen to you and should convince you that the best place for the approach plate is right straight in front of you.)

By the numbers, here is the approach study method:

1. Clean up the cockpit.
2. Find out what approach is in use and place that chart where it can be easily seen.
3. Memorize the missed approach procedure (at least the first *heading* and *altitude*).
4. Convince yourself that you will have to make a missed approach.
5. Orient yourself to the airport layout as it will appear when you break out of the clouds.
6. Study the entire approach—the "big picture."
7. Set up the radio gear for the approach.
8. Memorize the DH or MDA, as appropriate.

Although it seldom happens, be ready for a missed approach every time. This includes a quick mental review of the mechanics of "going around"—for most airplanes, this means power up, pitch up (to climb attitude), flaps up, gear up—as well as an early decision concerning your intentions immediately thereafter. Remember that controllers have a standard response to a missed approach report: "Roger, 34 Alpha, what are your intentions?" Compose your answer well ahead of time, *every* time you start an approach, and on that rare occasion when you need it, the words will come flowing forth just as if you know what you're doing.

# ✚ Wind-Shear Problems

It's now accepted that when an airplane flies through a wind shear, performance is very much affected. Consider an extreme example: An airplane flying at 100 knots in a 100-knot headwind would have zero groundspeed; if the wind died suddenly, the momentum of the airplane would carry it forward, but the airspeed and groundspeed (they're now the same) would have to instantly increase 100 knots if the original conditions are to be maintained. The heavier the airplane and/or the greater the initial airspeed, the more time is required for this acceleration to take place. (A rapid downwind turn may cause the same thing to happen.)

Whatever the span of time involved, there will be a nose-down pitch change and a net loss of altitude as the pilot tries to get the airplane back to where it was. Combine the high momentum of a heavy, fast airplane with the relatively long period of time required for a jet engine to "spool up" from a low power setting, then set up a decreasing-headwind shear at low altitude, and the classic undershoot accident often results.

Fortunately, low momentum (light weight, low speed) and the rapid acceleration characteristic of recip-powered aircraft take the bite out of all but the worst wind-shear situations for the lightplane community. However, the possibility is always there, and every now and then someone manages to ignore the indications and plants a lightplane in the approach lights or in a cornfield a mile short of the runway. Wind shear is worth knowing about and worth getting ready for no matter what kind of airplane you're flying.

It's true that wind shear can't be seen, but there is meteorological information available that should make knowledgeable pilots alert for the possibility of an encounter. First, whenever you've been fighting a headwind (or enjoying a tailwind) enroute, but find out that the destination airport's surface winds are calm, look out because, somewhere between your present altitude and the ground, the wind is dying. Second, whenever a cold front moving at 30 knots or more will arrive at the airport the same time you do, expect significant wind shear as you fly through the frontal surface. Third, a warm front with a temperature difference of only 10° may present wind-shear problems. And fourth, expect serious wind-shear encounters *anytime* there is a thunderstorm in the immediate environs (perhaps within 10 miles) of the airport.

Should you become aware of these conditions as you prepare for an instrument arrival (or a visual one, for that matter), plan to fly the final approach faster and higher than normal. And most important, if airspeed starts to bleed off, don't hesitate to add power right now; there's nothing that will solve the problem more rapidly or positively.

Wind shear will become most noticeable on an ILS procedure, when you are attempting to descend on a fixed, electronic glide slope. If you've done your homework so that you are familiar with the airplane and know the power setting that will keep the glide slope needle centered during a normal approach, the power gauges will tell you right away that either a strong tailwind or headwind is working; more power required means more headwind, and vice versa. In either case, prepare yourself mentally for some interesting possibilities as you get close to the ground. All the warning flags should fly when you recognize the need to carry more power than normal to stay on the glide slope at the same time as the Tower is reporting the surface wind as calm.

The tailwind-to-headwind shear situation (or a sudden reduction of tailwind) produces the opposite effects on the airplane; airspeed will increase, and the nose will tend to pitch up. Your instinctive response would be to reduce power and lower the nose, a combination guaranteed to produce a higher sink rate than before. If the wind subsequently drops off at an altitude too close to the ground to allow recovery, an unplanned touchdown may well occur. Whatever type of wind shear you encounter, the procedures should be the same; make yourself aware of the situation, carry extra airspeed and altitude (within reason), and at the first hint of a problem, power up and stay high.

# ✛ Visual Aids to an Instrument Approach

There's a surefire way to get into serious trouble at the end of an instrument approach procedure: Look up from the instruments as soon as you see something on the ground and continue trying to fly the airplane with visual clues only. That sort of piloting technique has probably wrapped up as many airplanes as any other single approach problem. We are taught from the very beginning to believe what we see, but at the crucial moment (at least from the point of view of an instrument pilot), our eyes play potentially hazardous tricks on us; it's important to understand the limitations of the human visual system under these conditions and to know what to expect from the various visual aids provided to help keep you out of the treetops during the last stages of an instrument procedure. Lighting systems and other visual aids discussed here are indicated on the approach charts; you should be aware of what's available before the approach begins.

General aviation pilots (particularly those who fly light airplanes) will frequently make approaches to airfields with minimum facilities; in this case, the only visual aid is the runway itself, augmented by lighting with a variety of configurations you won't believe until you've been there. Whatever the situation, it's important to realize that the human visual system is not designed to shift rapidly from one set of clues (the instruments) to another (what you see through the windshield) and consistently and properly interpret what's really there. So, a procedure must be developed; when the runway (or runway lights) becomes visible, begin to include those clues in your cross-check. Treat what you see out front as just another instrument indication, but give the outside clues an increasingly larger share of your scan time until the transition is complete. It only takes a few seconds, but doing it this way eliminates the possibility of making control inputs based on what you *think* you see. Even when the visibility is only half a mile, you will find that you can effect the transition from all-instrument to all-outside clues quite smoothly in the 15 seconds or so between "runway in sight" and touchdown.

Various types of approach lighting systems (ALS) have been developed and installed throughout the country—most ILS runways will be so equipped—and you'll find that the light pattern, colors, distance from first light to threshold, and intensity of these visual aids may differ considerably from one airport to the next. As a result, you may "see" things differently when looking at one type of installation as opposed to another, even though you are located at the same point on the approach. The key to success in this situation is to recognize that there probably will be some visual illusions and that they *must* be consciously overridden by information from the aircraft instruments such as airspeed, altimeter, attitude indicator, and so on. The lights are intended to: (1) make the runway environment visible as soon as possible in low-visibility conditions, (2) provide information relative to lateral displacement from the runway centerline, (3) provide information relative to the roll attitude of your aircraft. *Notice that there is no altitude information available from an approach lighting system per se;* that's why you have an altimeter. When in doubt, add at least 50 feet to the airport elevation and resolve not to go below that number on your altimeter until there's concrete underneath the airplane.

A high-time professional pilot who had been operating in the Pacific Northwest for some years was called upon to pick up his employer late one winter night at a small, ILS-equipped airport. The cloud ceiling was very close to the Decision Height of 200 feet, but visibility was quite good under the clouds.

The ALS was in operation, and as far as could be determined in a postaccident flight check, the localizer and glide slope signals were normal in every respect. Yet, for reasons known only to the pilot, the airplane struck the ground almost

2 miles from the runway threshold. The wings were level, gear was down, and flaps were set to the position normally used during an instrument approach procedure in this airplane, suggesting that the pilot was in complete control at impact.

The most likely explanation of this accident centers on the pilot's perceived need to land (his employer was known to be a very demanding person, intolerant of late arrivals, to say nothing of his pilot not showing up at all) and the illusions probably produced when the airplane broke out of the low clouds. If the pilot had indeed consciously flown below the glide slope in an effort to guarantee "breakout" and landing, he would surely have noticed that the ALS appeared much flatter and smaller (that is, farther away) than would be expected during the final segment of a normal approach to minimums.

Even if the glide slope signal or the onboard equipment had malfunctioned, the old reliable altimeter would have saved the day for this pilot. By making a "no exceptions" decision to fly no lower than the published Decision Height, the accident could have been prevented. Of course, the boss might have thrown a fit, but the pilot would have been alive to quit his job...instead of losing it.

One of the toughest judgment calls in instrument operations is deciding when to commence descent to the runway on a nonprecision (no glide slope) approach after the airfield has been sighted. With no lighting systems available, you'll be glad you practiced ahead of time and know the power setting and pitch attitude for a normal descent to a landing. Using this as a base, you can start down, using power to correct the glide path when you sense the need; above all, don't go below your predetermined "minimum-minimum altitude" (runway elevation plus 50 feet) until the threshold is behind you.

There is another type of visual aid so simple that you've got to wonder why no one thought of it when the very first instrument approach procedure was developed. It's nothing more than a radio marker beacon or a DME fix at the point where a 3-degree glide slope intersects the Minimum Descent Altitude. This marks the point at which you should begin a normal descent to the runway, *if* you have established visual contact. The VOR-DME approach in Fig. 14-1 is typical, with the Visual Descent Point (VDP) shown in the profile view; when on final approach and at MDA, you should start down when the DME counter indicates 3.0 miles. Once again, it's extremely helpful to know ahead of time the approximate power setting needed to maintain a safe descent path.

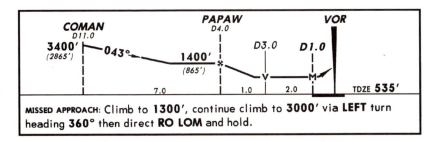

**Fig. 14-1.** *Visual Descent Point (VDP) shown on Jeppesen approach procedure chart. (Copyright Jeppesen Sanderson Inc., 1985, 1986; all rights reserved; not to be used for navigation.)*

There are two possible cases when using this visual aid: (1) arriving at the VDP with no runway in sight or (2) sighting the runway before the VDP is reached. In the first instance, get ready for a missed approach, especially in a high-performance airplane. If you haven't spotted the runway by the time you're this close, there probably won't be time for a safe descent when the airfield does come into clear view (as you continue beyond the VDP in level flight, the required angle of descent gets progressively steeper). In the second situation, fly to the VDP before leaving the Minimum Descent Altitude; an early descent will produce a shallow glide path that's filled with illusions. Why upset your visual apple cart at such a crucial time? Fly to the Visual Descent Point and make a *normal* approach to the runway.

The "when to descend?" and "how fast to descend?" guesswork is eliminated by the installation of a Visual Approach Slope Indicator (VASI; pronounced "vassy"), which consists of two light sources just off the side of the runway near the approach end. (Figure 14-2 is a side-view schematic of a VASI installation.) When you break out of the clouds at MDA, you'll most likely see two red lights indicating that you have not yet flown into the predetermined glide slope. As you continue the approach at MDA, the nearer light will gradually change to white, at which time you should power down and start your descent to the runway. If you adjust power to keep that color combination— red over white—in view, you will be flying along a glide slope guaranteed to bring you safely across the runway threshold. The VASI glide slope is typically elevated about 3 degrees, the same as most ILS glide slopes. Should your rate of descent take you above the glide slope, you'll see two white lights; if you go below (that's a no-no!), the lights will turn red. Most important to IFR pilots, don't commence the descent unless and until you see a clear red-over-white indication from the VASI lights. At some large airports served by wide-body

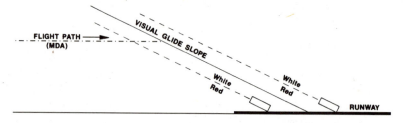

**Fig. 14-2.** *The Visual Approach Slope Indicator (VASI). Optical filters project white light above the lamps' centerlines, red light below.*

jumbo jets, you may see *three* sets of VASI lights alongside the primary runway, providing two glide paths: an upper one for the big airplanes and a lower one for the little guys. The pilot's job is to fly one or the other. (To avoid wake turbulence when you're following a large aircraft, stay on the *upper* glide path. Interpret the lights in the same way: red over white means "you're all right.")

The VASI system just described will usually be installed on the primary instrument runways at terminals served by air carriers and at most tower-controlled airports. If you fly frequently to smaller, non-towered airports, you should expect to find other versions of visual indicators. There are three-color systems (red for low, green for on the glide slope, amber for almost on) and pulsing-light systems that pulse faster as you depart from the glide path. On occasion, you'll come across the most basic system of all: three rectangular "signboards" beside the runway that will be lined up horizontally only when you are lined up vertically with the proper glide slope.

Whatever system you encounter, remember that you are required to stay on the glide slope until it's necessary to go lower in order to land. Well, that's not quite true. The FARs speak only to VASIs installed at tower-controlled airports, but in the interest of safety, it's a good idea to stay on the glide slope anytime one is provided.

# 15

# Instrument Approach Procedures

From a "do what you're told" radar approach at the low end of the simplicity scale to the Instrument Landing System where the entire job of navigation is up to you, every instrument approach procedure is just a means to an end. When enroute navigation has brought you through the clouds to that fabled "point B," the approach charts take over to guide you to a point from which you should be able to land the airplane—it's that simple! Whenever the navaids provide a more accurate position on the final approach segment, that point gets closer to the runway, horizontally and/or vertically. It will be at its maximum distance on a typical NDB (ADF) approach (because of the inherent inaccuracy of the system and the great amount of pilot navigational inputs required), moves closer with VOR/RNAV/GPS, gets even better on a localizer approach, and gets down very close to the runway with the most precise procedure of all, the ILS.

You may have personal opinions about the accuracy of the various types of approaches, but when it comes to precision, there is a major classification already laid down for you; all approaches are put into one of two general categories: precision or nonprecision. (The semantics can be argued because some of the so-called nonprecision procedures will lead you right down the paint stripe in the middle of the runway, but there are important implications stemming from these descriptions, which are basic to a complete understanding of instrument approaches.) If the installation provides electronic glide slope information, it's known as a precision approach, and any procedure without a glide slope falls into the nonprecision category. For a typical approach procedure, see Fig. 15-1.

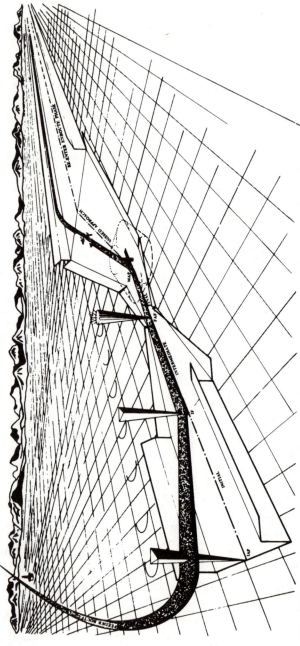

**Fig. 15-1.** *Segments of a typical approach procedure, showing how terrain and obstacle clearance "funnels down" toward the runway.*

For all practical purposes, the only precision approach in use today is the Instrument Landing System (ILS), in which the glide slope is displayed in the form of a moving pointer or indicator of some sort that enables flight path adjustments to conform to a prescribed electronic path leading to the runway.

Military Precision Approach Radar (PAR)—also known as Ground Controlled Approach (GCA)—qualifies as a precision procedure, but these facilities are available to civilian pilots at only a handful of joint-use airports in the United States. Using this system, your position relative to the PAR centerline and glide slope are observed on superaccurate radar, and corrections are made by the pilot at the controller's direction. This means that the approach can be accomplished with only a radio receiver and basic instruments operating in the aircraft. It could be a life-saver if your navigation radios fail in down-to-minimums weather and you happen to be within range of one of these military fields.

# ✢ How Many Black Boxes Do You Need?

The least complicated approach procedure would be a radar approach, for which you need only some means of receiving voice communication. For a VOR or GPS procedure, at least one receiver is required, and an NDB approach cannot be accomplished without an ADF on board. To get to the Decision Height on an ILS approach, you must be able to receive the localizer and the glide slope. (The two marker beacons provide valuable range information, but the law allows you to substitute radar or your ADF for the outer marker in most situations.)

Some VOR approach procedures include DME fixes that allow you to position yourself more accurately on final approach, and lower Minimum Descent Altitudes are therefore authorized. If you don't have DME or if it's not working, you're stuck with a higher MDA. On the other hand, an approach procedure that is specifically titled "VOR-DME" requires *both* receivers; if either is inoperative, you're out of business.

A few procedures require more than the basic navigational aids because of local conditions. They fall into two groups: those that *absolutely require* additional equipment and those in which lower minimums are *available* for aircraft with more radio gear (or when radar service is available). At Cincinnati, Ohio (Fig. 15-2), radar and ADF are required for the approach, because you must be vectored to MAMMY, and the missed-approach holding pattern is based on a nondirectional beacon.

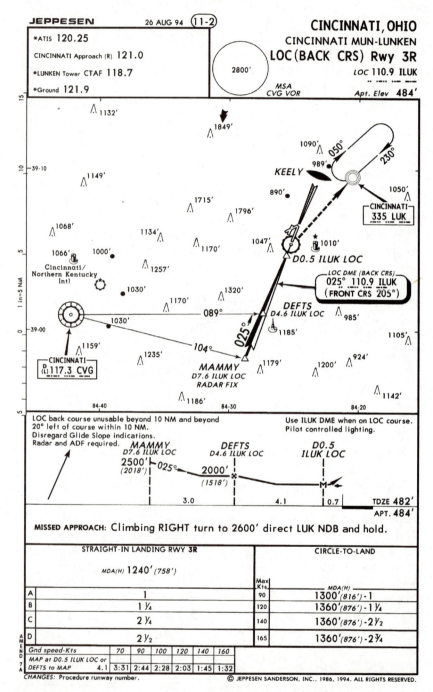

**Fig. 15-2.** *Approach procedure chart for Cincinnati, Ohio (Copyright Jeppesen Sanderson Inc., 1986, 1994; all rights reserved; not to be used for navigation)*

The VOR Runway 27 procedure at Lakeland, Florida (Figure 15-3) can be executed with just a VOR or GPS receiver, but you may descend 360 feet lower if you can determine the 4.0-mile point; this requires dual VOR or VOR and DME. On a low-visibility day at Lakeland that 360 feet could be the difference between a successful approach and a go-around.

With radar approach control, more often than not you will be given a vector to the final approach course, and so the procedure turn is slowly going the way of the position report. But you must know how to fly the complete approach as published because there may come a day when radar or communication radios fail and you'll have to fly the whole procedure yourself.

Once you've learned the basic segments of an approach procedure, and the principles underlying pilot actions at various times, you can fly any approach in the book. Each procedure is discussed using this format:

Description

Setting up the radios

Initial approach

Station passage and tracking outbound

Procedure turn

Final approach

Straight-in approaches (those that lead directly to the landing runway with no maneuvering) are the first order of business, followed by circling approaches, and the next logical step when the weather is lousy, the missed approach.

An instrument approach doesn't have to follow the printed directions to the letter, but if you're going to take advantage of the legal shortcuts, you must know what you're doing. Pilots who accept a contact approach but have no idea what they are expected to do can become hazards in the airspace. So contact and visual approaches are discussed, followed by radar approaches and several other procedures you should know about to make the most of your IFR capabilities.

Pilot complacency, controller error, and hostile terrain make a dangerous mixture. Such was the case in a West Coast episode involving an airport that lay at the end of a long, narrow valley. Although this ILS isn't a "thread-the-needle" approach, there are ridges on either side that rise well above the initial approach altitude.

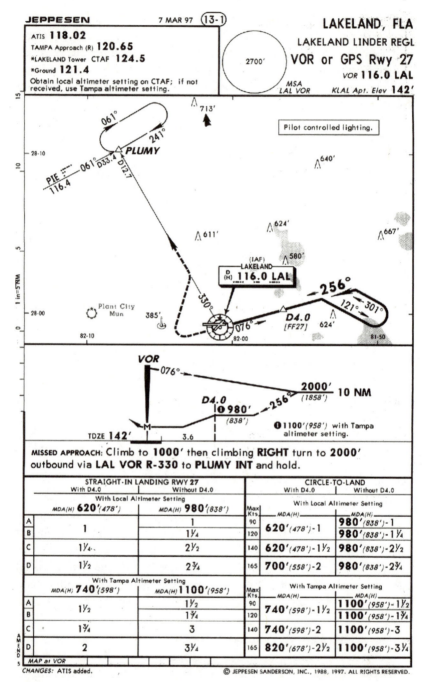

**Fig. 15-3.** *Approach procedure chart for Runway 27 at Lakeland, Florida. (Copyright Jeppesen Sanderson Inc., 1988, 1997; all rights reserved; not to be used for navigation)*

The pilot had flown the ILS on many occasions VFR and IFR—because of his company's frequent trips between this airport and others in that part of northern California. This flight seemed completely routine on all counts, with a ceiling of several hundred feet and good visibility under the clouds. Unfortunately, as the flight approached the airport, the approach controller became confused during a hand-off from his Center counterpart and commenced vectoring a blip that was *not* the airplane intending to use the ILS procedure.

Assuming that the pilot of the ill-fated airplane had set up his #1 nav indicator for the ILS when the vectors began, he would have noticed that he was being "vectored" well to the right of the localizer instead of to the left, which was normal and which *appeared* normal to the controller, who was unfortunately looking at another target. When the vector was issued to turn the flight around for localizer intercept, the pilot should have noticed that it was a turn *away from the localizer* and toward terrain higher than his present altitude.

In due time, the controller noticed that the blip was not responding to his vectors and inquired of the pilot if he were in fact turning to the new heading. About this time, the pilot also sensed that something was not right and questioned the vector, but it was too late; even a full-power go-around wasn't enough to clear the ridge and the plane impacted short of the top, killing all aboard.

As always, there's something to be learned from this accident. Study each approach beforehand with a thought to what *should* happen, what the navigational indications *should* be, and stay on top of your actual position throughout the procedure, particularly during that portion of the approach that consists of radar vectors. This is especially important in those parts of the country where terrain and/or obstacles are significant. It's called "positional awareness," and it provides the foundation on which you can decide whether the controller's instructions—or your own maneuvering—are proper, safe, and adequate to get the job done.

And whenever you have even a shred of doubt about the outcome of a vector, turn, or descent, question yourself or the controller. One quick inquiry might save the day.

# ✝ **Practice Is a *Must*!**

Pilots aren't the only egomaniacs in the world. A famous musician once said, "If I practice four hours a day I am the world's greatest pianist; if

I practice *eight* hours a day, I am Paderewski!" Practice improves performance in any endeavor, but it becomes especially significant when you're talking about instrument flying skills.

Practicing the approach procedures published for your home airport is commendable, since you'll probably make most of your down-to-minimums approaches there, but this kind of practice doesn't do much for your versatility. The time that you set aside for instrument practice (and you should do this unless you fly IFR regularly and frequently) can be more productive if you will have your instructor or safety pilot pick an instrument approach procedure at random from the chart book, give yourself a few moments to study it using the method in Chap. 14, and then fly it. Of course, you can't practice ILS or localizer approaches this way, but if you can receive a signal from any VOR or NDB, you can use that signal to simulate an approach procedure; a local AM broadcast station is often convenient for ADF practice.

To make these exercises as realistic as possible, reset your altimeter to an indication that will let you fly the published altitudes and put the MDA for the approach at a safe level. Under the hood, you should be concerned only with the numbers you see on the altimeter. It makes no difference to you what your actual altitude is or where the navigation signals are coming from; keeping you clear of things is your safety pilot's job. After a few of these sessions with randomly selected procedures, you'll have convinced yourself that all approaches are generally the same; there are different altitudes, different headings, but you won't be surprised when cleared for a procedure you've never seen before.

Instrument approaches are the command performances of the flying business; the ones you make when the weather is right down on the deck are the ones you practice for, so when you have the opportunity, get under the hood and really *work*. Don't bore expensive holes in the sky.

# ✝ Procedure Turns

It's refreshing, in this heavily regulated aeronautical world, to have one IFR procedure that you can execute just about any way you please. To be sure, there are limitations on the airspace you may use for a procedure turn, but by and large, how you reverse course during an instrument approach is up to you. A procedure turn does nothing more than get you turned around and headed back the way you came, on a specific course.

Three types of procedure turns have evolved over the years, and they are best classified in terms of their relative difficulty. (Not to imply that any one is actually harder to perform than the others, but the elapsed time from start to finish grows shorter as you move up the scale, and this means that you have to be very much on top of the entire situation when using the most rapid reversal.) The "old standard" is the 60-second procedure turn, and it requires at least 3 minutes from the time you leave the outbound course until you're back to where you started the maneuver. The 40-second type (really just a modification of the 60-second turn) cuts the total time somewhat depending on the wind, and the fastest one of all is the 90-270; you'd better be ready for your inbound chores shortly after reversing direction. (Standard-rate turns are assumed in each case.)

Procedure turns have several things in common: They must be executed at or above a specified minimum altitude (although *maximum* and *mandatory* procedure turn altitudes show up every now and then); you may not go beyond a specified limit from the approach facility as you reverse course; and your maneuvering must be accomplished on the specified side of the course. The restrictions are designed to allow plenty of room, yet keep you clear of obstacles as you turn.

The maneuvering side of the course (indicated by a barbed arrow on the U.S. charts or Jeppesen's full depiction of the turn) and the procedure turn altitude (shown on the profile view of the approach procedure) are quite variable, but the maximum distance is 10 miles from the approach fix on almost every chart. And therein lies a trap for unwary pilots who always allow themselves the full 10 miles; some approaches cannot guarantee terrain or obstruction clearance for this distance, and a shorter mileage is quoted. "Nonstandard" procedure turn distances are not emphasized on approach charts, and it's easy to count on 10 miles, only to find yourself in trouble. Pay close attention to the maximum procedure turn distance on *every* approach, with airports in the mountainous sections of the country most suspect.

All three types of turns are safe procedures if you remain at or above the procedure turn altitude *until established on the approach course inbound,* when you may descend to the next lower altitude, if the published procedure permits.

## The 60-Second Procedure Turn

Used by some pilots because of its simplicity and the fact that it gives you a bit of a breather on the inbound leg, the 60-second procedure (Fig. 15-4) is commenced by a standard-rate 45-degree turn away

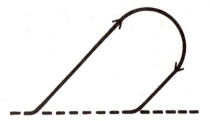

**15-4.** *The 60-second procedure
turn.*

from the outbound course (headings are given on the chart) and then
flying for 1 minute on the new heading. When the minute is up, turn
180 degrees *in the direction opposite the first turn,* or always turn *away
from the airport,* whichever is easiest for you to remember. In a
no-wind situation, you'll have about half a minute to fly on this heading before you reintercept the inbound course. It's a good time to
refresh yourself on the MDA or DH, your missed approach procedures,
or your anticipated circling maneuvers, depending on the kind of
approach you are flying. When your navigation displays show that
you're approaching the inbound course, lead the turn a bit and track to
the fix.

   The only timing required is on the 1-minute leg outbound; start
the time as soon as you level the wings after the 45-degree turn,
assuming that you're on course when you start the procedure turn.
This rule also holds for those times when you find yourself *slightly* off
course on the maneuvering side; start timing when you level the
wings. However, should you begin the maneuver from the *other* side
of the course, don't start timing until your instruments indicate that
you have crossed the course.

   In a situation where you have no idea of the wind conditions, the
60-second procedure turn will probably give you the best break; if
there's no wind, it will show up as a normal 30-second trip from completion of the inbound turn to the final approach course. If there's a tailwind outbound, you may need a box lunch to survive the journey back
to course. And in the worst situation—a strong headwind outbound—
the navaids will indicate course interception during the inbound turn;
in this case, turn directly to the inbound heading.

   Safe, simple, and sure, the 60-second reversal is a good one to use
when in doubt. But don't get the idea that it's just for beginners and
those who have time to spare. There are occasions when it becomes
necessary to lose a lot of altitude during a procedure turn, and that full
minute outbound, another in the turn, and part of a minute before

intercepting the inbound course can mean a couple thousand feet, even at a moderate rate of descent.

## The 40-Second Procedure Turn

Here's a way to make the procedure turn more efficient by controlling the effects of the wind instead of accepting whatever displacement it may cause. It's a modification of the turn just discussed and involves changes in the time outbound as well as the point at which you will reintercept the course. The 40-second turn (Fig. 15-5) will obviously cut down the time you spend reversing course and can be your contribution to speeding up the flow of traffic during rush hour at the local airpatch. But it requires a little more of the pilot, since you must calculate roughly the effect of wind, apply the correction, and perhaps more important, be prepared to cope with the faster return to course that results. There'll be no breather after you turn inbound on this one; if properly done, you will roll out *on course,* and you must be all set to proceed with the remainder of the approach.

The 40-second turn begins the same as its 60-second kin, with a 45-degree turn away from the outbound course, and timing also starts at the same place. But the duplication stops right there because now you will add to or subtract from the 40-second base figure a time factor to correct for wind. The calculation puts together the number of degrees of wind correction that was required to keep you on the outbound course and in which direction. If your procedure turn takes you downwind, *subtract* 1 second for each degree of crab required on the outbound course; when turning into the wind, *add* 1 second for each degree. Thus, you will shorten or lengthen the first leg and change the radius of turn as you head back for the inbound course. If this is done properly (and "properly" includes all turns made at the same rate), you should find the appropriate needles centering as you roll out on the inbound heading, plus or minus drift correction, of course.

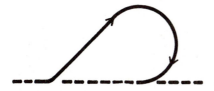

**Fig. 15-5.** *The 40-second procedure turn.*

A well-executed 40-second procedure turn should conclude with a smooth, sweeping turn onto the approach course, and even if you are off a few seconds in your calculations, you will have saved some time. This is also a good one to use when there is no wind and you wouldn't mind getting on the ground a little sooner. It's a difference of only half a minute or so, but when the terminal area is saturated with IFR traffic making procedure turns, those half-minutes add up.

## The 90-270 Procedure Turn (or 80-260, If You Prefer)

Maybe it all started with an IFR pilot not being able to make a decision after the first 90 degrees of turn. Whatever the origin, the 90-270 (Fig. 15-6) will get you turned around in the minimum time and with the minimum outbound displacement. You will also find it very helpful as a low-visibility circling approach when you need to fly the length of the runway and then reverse course to land. There is no timing involved, and it will work in all but the strongest winds. The key is to make your turns at a uniform rate and to change heading a *full* 90 degrees on the first turn.

Choose carefully the point from which you begin a 90-270 because you will come back onto the inbound course not very far from where you left it. If you have more than 1000 feet to lose in the turn, you're going to be plunging earthward at a rapid rate, which will speed up the whole approach to the point where things may be happening so fast you can't keep up. Having to execute a missed approach because you started a rip-snorting 90-270 too close to the fix is hardly a laudable demonstration of efficiency.

When the distance out and altitude-to-lose are compatible, start with a 90-degree turn in the appropriate direction and, *immediately* on reaching the 90-degree point, roll smoothly into an identical angle of bank in the opposite direction. Depending on the wind condition,

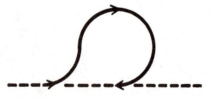

**Fig. 15-6.** *The 90-270 (or 80-260) procedure turn.*

you will probably begin to center the nav needles about 45 degrees from the inbound heading; keep turning, and as you roll out, you'll be "right on." Remember that the accuracy of this maneuver depends on uniform, constant rates of turn and a 90-degree heading change at the outset. If you're holding 15 degrees of crab as you proceed outbound before starting the procedure turn, count 90 degrees from that heading, not from the course you are on. (Some pilots like heading changes of 80 degrees and 260 degrees; it's a matter of personal preference.)

## Outbound Timing

No matter which of these three procedures you elect to use, you must decide how far outbound you will fly before beginning the procedure turn. Bearing on your decision will be the maximum distance allowed, the amount of altitude you have to lose, and how much time you figure you will need (or would like to have) to get squared away on the final approach course. When the weather is hovering around minimums and you want everything going for you, give yourself plenty of time. Within reason, the distance you fly outbound before beginning the procedure turn should increase as the ceiling and visibility decrease.

A generally accepted figure for lightplanes is 2 minutes, which will supply that comfortable time pad (unless there's a terrific headwind outbound), yet keep you well within the mileage limit. Of course, you should slow down as soon as you cross the fix outbound. There's no reason for high airspeeds during a procedure turn; you'll use up more fuel, increase your radius of turn, and fly farther away from the fix. At a reasonable airspeed and 2 minutes outbound, you needn't worry about a 10-mile limit; a groundspeed of even 180 knots will keep you inside the envelope of airspace reserved for you as you maneuver.

## No Procedure Turn, Please

There are two situations frequently encountered which preclude the use of a procedure turn. The first occurs when you are issued a radar vector that the controller states will take you to the final approach course for the appropriate instrument procedure. The second involves an approach clearance via a route on the approach chart that bears the notation "NoPT," meaning of course "No Procedure Turn" (Fig. 15-7). In this case, the routing will result in your intercepting the final approach course in time to line up and descend if required before reaching the Final Approach Fix.

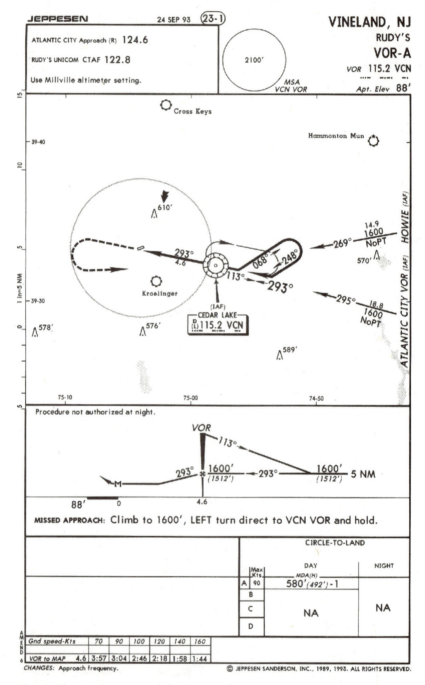

**Fig. 15-7.** *Initial approach routes that do not require a procedure turn (NoPT). (Copyright Jeppesen Sanderson Inc., 1989, 1993; all rights reserved; not to be used for navigation)*

# ✝ A Procedure Turn Is Not the Only Way

When is a procedure turn *not* a procedure turn? When it is omitted from the published approach or replaced by some other maneuver, as at Parkersburg, West Virginia (Fig. 15-8), where DME arcs and a lead-in heading from MORAN intersection provide adequate guidance to get lined up on the localizer course. On occasion, a holding pattern is substituted for a procedure turn to limit the outbound distance or to guarantee that pilots turning around will be doing so in a prescribed manner, instead of the several options available for a regular procedure turn.

One of the smoothest, least complicated procedures is the teardrop turn. You'll find one leading up to the ILS approach to Runway 31 at Dothan, Alabama (Fig. 15-9). Over the wiregrass VOR at cruise altitude and cleared for the approach, track outbound on the 155 radial, descend to not less than 2200 feet, and when you figure you can turn onto the localizer comfortably, do it. If you want to cut the elapsed time right down to the bone, watch the DME readout during the teardrop turn, and you can put yourself on the localizer as close to the outer marker as you want to be.

Another variation of the procedure turn is well illustrated by the VOR approach to Runway 20 at Fitchburg, Massachusetts (Fig. 15-10). You may be cleared to hold north of the NDB, waiting your turn for an approach. When you are cleared for the approach, should you execute a procedure turn? Take another look at the chart and notice that the missed approach holding pattern uses the same airspace as the charted procedure turn.

The only restriction you must observe here is the maximum distance of 10 miles, so be sure you stay within that limit, descend to procedure turn altitude during your final circuit of the holding pattern, and go ahead with the approach. The approach controllers would probably be unhappy if they observed you turn over the station, proceeding outbound on the 021° course and into a beautiful procedure turn. No matter how well executed, this maneuver would be unnecessary, ill-advised, and rather inconsiderate to the pilots behind you. Whenever possible and practical, continue in the holding pattern when cleared for an approach *if the holding pattern and procedure turn are coincident on the chart*. If time versus altitude loss becomes a problem, you can always extend the outbound leg of the holding pattern for 2 minutes or so before turning inbound.

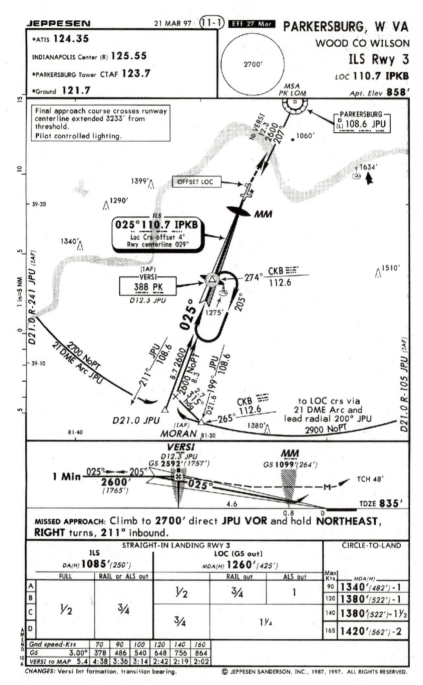

**Fig. 15-8.** *DME arcs and lead-in heading, which take the place of a procedure turn. (Copyright Jeppesen Sanderson Inc., 1987, 1997; all rights reserved; not to be used for navigation)*

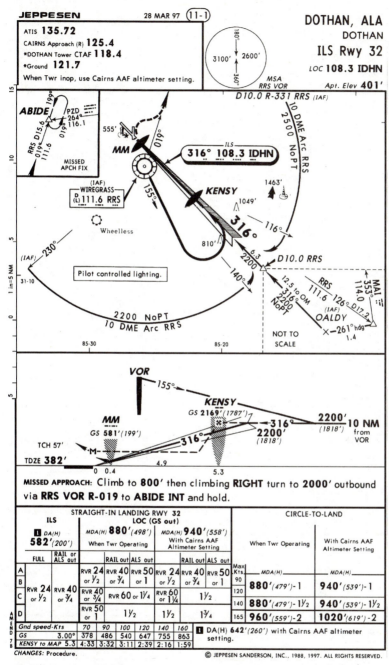

**Fig. 15-9.** *Approach procedure chart for Runway 31 at Dothan, Alabama, showing the teardrop turn in lieu of a procedure turn. (Copyright Jeppesen Sanderson Inc., 1988, 1997; all rights reserved; not to be used for navigation)*

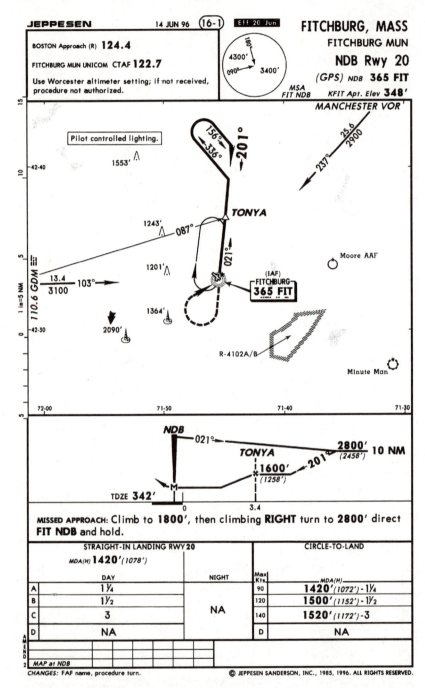

**Fig. 15-10.** VOR approach procedure for Runway 24 at Fitchburg, Massachusetts. (Copyright Jeppesen Sanderson Inc., 1985, 1996; all rights reserved; not to be used for navigation)

# ✞ Approaches to Nontowered Airports

When you are cleared for an approach to an airport that has an operating control tower (Class C or D airspace), you are guaranteed separation from other IFR traffic right down to the surface within 10 miles of the airport, plus extensions to accommodate published IFR procedures. Such airspace is usually designated as a "surface area," which means that the protection continues during the hours when the tower is closed (check the IFR charts to be sure).

In the case of nontowered airports located in Class E (uncontrolled) airspace, IFR separation is guaranteed only if the airspace around the airport is designated as a "surface area." Otherwise, a transition area is usually set up, but it doesn't do a complete job. Instead of protection all the way to the ground throughout the approach, you are in controlled airspace only during the time you are 700 feet or more above the ground (occasionally 1200 AGL; check the charts). This means that you can be cleared for an approach, fly the procedure exactly as published, and wind up in uncontrolled airspace. When you pass through 700 or 1200 feet AGL, you have gone from one world to another—from the *complete* protection of ATC to *no* protection except the CTAF and your own eyeballs.

This is not a fault of the system but is the sort of thing you should know about so that you can increase your vigilance. Just because you have been cleared for the approach, don't get the mistaken idea that you are the only one who is allowed to be there. In Class G airspace, the only requirements to fly VFR are 1-mile visibility and clear of clouds, so there is nothing to prevent some dedicated pilot from going out to the airport for a few touch-and-gos even though the weather is perfectly horrible. And it's not just the VFR types you have to worry about because there could be an intrepid pilot approaching the same airport IFR, without a flight plan, without a clearance, and with no notification to ATC. That is completely legal as long as the pilot remains clear of controlled airspace; not necessarily smart, but it's legal.

There are three ways you can lessen the danger when inbound to a nontowered, nonsurface area airport in IFR conditions. First, ask the controller to check for and keep you posted on radar targets in the vicinity of the airport. Second, get on the CTAF and broadcast your position and intentions as soon as you can and continue your broadcasts in the blind as often as you think necessary. Third, get down to the MDA early and keep your eyeballs moving. Remember that the minute you descend below 700 or 1200 feet AGL you must share

Class G airspace with whomever else may be using it; you're in the arena with the scud runners and obligated to see and avoid, just as you hope they are doing.

# ✝ Instrument Procedures with No Final Approach Fix

The Maltese Cross symbol that appears in the profile view of most approach charts indicates the Final Approach Fix (FAF). It's the last electronic fix before beginning the descent to MDA and the point at which you should begin timing for the missed approach point. (The FAF is indicated on an ILS procedure as well, and it's intended for the localizer-only approach; when flying a full ILS, start descending when the glide slope needle centers.)

But if you riffle through the book of approach charts, you'll find a number of procedures *without* a Final Approach Fix—no Maltese Cross. In these cases, the radio aid (VOR or NDB) will usually be located right in the middle of the airport; you'll also find that the time to fly from FAF to MAP is missing. With no point from which to start the time, there's no choice but to fly at MDA until station passage, which is not all bad because you're flying toward an increasingly accurate signal and the missed approach point is un-mistakably clear.

Without timing, radar service, or DME to provide you with dis-tance information, the no-FAF procedure has a built-in trap in that the outbound leg (on the way to the procedure turn) starts over the air-port, not over a beacon or VOR a mile or three from the runway. Pilots who fly a normal 2-minutes outbound may wind up still letting down to MDA when they roar across the airport or else descending at a very rapid, unsafe rate when the problem is recognized.

The solution is simple: Ask for radar distance information; set up your DME to provide it if the station is reasonably well lined up with the final approach course, use a nearby VOR radial to let you know how far to fly before turning around, or put an RNAV waypoint on the airport. In the absence of any of these, extend the outbound timing so you'll have room to descend on the way back to the airport. And as soon as it's legal—that is, on course, inbound—head for the MDA; don't hesitate. The idea is to reach MDA as soon as you can do it safely. Grinding along at MDA for an extra half-minute or so is not sinful; besides, it gives you better odds on breaking out of the clouds and more time in level flight to look for the airport and make a ratio-nal decision about when to start down for the landing.

# ✝ Circling Approaches

"Barnburner 1234 Alpha is cleared for the Runway 28 ILS approach, circle to land Runway 14." If you operate into airports where only one runway is served by an approach procedure, you can bet that you'll hear clearances similar to that one many times, since the wind doesn't always favor the primary runway. A circling approach increases the flexibility and utility of the airport, but like anything else, it costs something; in this case, the price is higher minimums and some rather stringent additional requirements to keep you out of trouble while you're circling.

The criteria that make an approach a circling approach are quite straightforward; whenever the landing runway is aligned more than 30 degrees from the final approach course or when a normal rate of descent from the minimum IFR altitude to the runway is considered impossible, a circling approach is established. Because either or both of these criteria cannot be met, some airports offer no straight-in minimums. This situation is indicated in the title of the approach chart; an alphabetical designator added to the title—such as NDB-A, VOR-B, or GPS-C—means circling approaches *only*.

You may be cleared for a circling approach at the end of *any* published procedure from NDB to ILS; the controller's motives are honorable, and you won't be asked to circle unless wind and weather conditions are such that landing straight-in would present a hazard. "Cleared to circle" does not mean that you are *required* to maneuver; if you have the active runway in sight in time to make a normal, straight-in landing, by all means do so, keeping the appropriate controller informed and, of course, obtaining the appropriate clearance.

No matter how your circling approach begins, it is going to turn into a nonprecision procedure; there's currently no way to provide a circling glide slope for some other runway. Even if you are cleared for an ILS approach with a circling maneuver on the end, *you may not go lower than the circling Minimum Descent Altitude*; in addition, you will find no circling approaches in which the required visibility is less than 1 mile.

The circling MDA guarantees at least 300 feet of vertical clearance from all obstacles in the circling area, which is the connected radii from the ends of all the airport's runways (Figure 15-11). When you are cleared for a circle-to-land maneuver, you should shift mental gears as you prepare for the approach, and when studying the chart, don't even look at the straight-in minimums. Let the circling MDA burn itself into your mind.

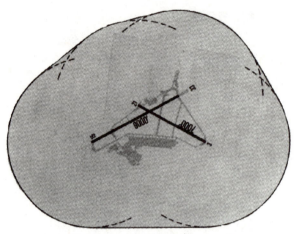

**Fig. 15-11.** *Circling area around a typical airport. You are guaranteed at least 300 feet of obstacle clearance within this area, the size of which varies with approach category.*

The size of the circling area is based on the amount of airspace that might be required for maneuvering to a landing on any one of the airport's runways. In recognition of the smaller space required for slower airplanes, the planners came up with a set of "Approach Categories," and the MDAs and visibility requirements for each are printed on the approach procedure charts (Table 15-1). Computation of the category for any airplane consists of determining the power-off stall speed in the landing configuration at maximum gross weight and multiplying that number by 1.3.

The legal requirements you must observe while "circling to land" in this situation are: (1) you must not descend below the circling MDA until descent is necessary for landing (practically, this means when

**Table 15-1.** Circling Approach Categories and Speed Limit

| Category | Speed |
|----------|-------|
| A | Less than 91 knots |
| B | 91–120 knots |
| C | 121–140 knots |
| D | 141–165 knots |
| E | 166 knots or more |

turning base or final for the landing runway); (2) you must keep the airport in sight throughout your circling maneuver; and (3) you must remain clear of all clouds. Should you lose sight of the airport, you are bound by law to execute an immediate missed approach.

Obviously, it would be unwise to begin circling before you have the field in sight, so fly the final approach course right down to the circling MDA and, if the field doesn't show up, go around. But suppose that you have eased down to the MDA well ahead of time and at 1 mile the airport begins to take shape ahead of you. If you have done a good job of studying the airport layout during the preparation for the approach, you will know just how it will look, and more important, you will have decided which way you are going to circle for landing. Always check the remarks section of the approach chart before commencing a circling approach; there may be some surprises waiting for you if you maneuver where you shouldn't.

How you get lined up with the runway is your business, as long as you comply with the restrictions that have been mentioned. There are several generally accepted patterns suggested for low-visibility approaches (Figure 15-12), and it's difficult to imagine a situation that would require a maneuver much different than one of these. The important thing is to decide way back at the beginning of the approach what you're going to do when the field comes into view and then stick to your plan; 1-mile visibility and only 400 or 500 feet of altitude doesn't give you much room to change your mind.

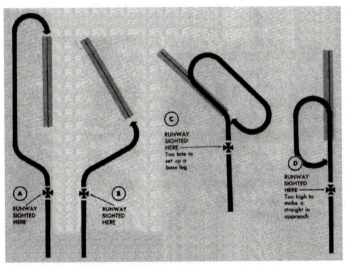

**Fig. 15-12.** *Suggested maneuvers for circling approaches in low-visibility conditions.*

The airport at Aspen, Colorado, is tucked into a valley that terminates rather abruptly in the mountains that make Aspen the skiing capital it is. And the terrain that makes for such good skiing also makes the instrument approach a bit dicey, to say the least. It's a "circling only" situation because an MDA high enough to clear the surrounding terrain makes it impossible to descend to the runway at a normal rate of descent, especially if you happen to be flying a very high-performance business jet.

That was the situation several years ago when the pilots of a Lear 35 started the approach into Aspen. The weather was typical of February in the mountains—snow showers moving across the field, otherwise not all that bad. (The reported visibility was 2 miles, 1 mile less than the published minimum. This crew was operating under Part 135, which prohibits the commencement of an approach unless the reported visibility is at or more than the minimum. They started the approach nevertheless. That's strike one.)

When the Lear broke out of the clouds at MDA and the pilot spotted the runway, the relatively wide, flat area to his left which usually provided plenty of room for circling was blocked out by a snow shower. The pilot elected to circle to the right, descending all the while, which forced him to fly behind a mountain ridge and lose sight of the airport. Strike two.

Still descending, the pilot apparently recognized a canyon perpendicular to the runway, and followed it to a major highway that parallels the centerline. But the low airspeed and the steep-banked attempt to get lined up with the runway caused the Lear to stall. It crashed well short of the threshold, killing all aboard. Strike three, you're out.

Part 91 of the FARs lays down the law for *all* pilots when it comes to circling approaches. You may not operate below MDA unless the aircraft is continuously in a position from which a descent to a landing on the intended runway can be made at a normal rate of descent using normal maneuvers. Part 91 also requires that an immediate missed approach be executed whenever an identifiable part of the airport is not distinctly visible to the pilot during a circling maneuver at or above MDA. (Losing sight of the airport because a wing or a windshield post gets in the way is okay; losing sight of the airport because of an intervening mountain is not.)

There's a lesson here: Pay attention to the rules.

Every approved instrument approach procedure includes a set of very specific instructions for your guidance in the event you are unable to see the runway at the missed approach point. This procedure is also to be followed when you abort a circling approach, but now you will find yourself someplace other than the straight-in missed approach point. There's a special rule for a circling missed approach that will keep you within the allotted airspace. It says that when executing a missed approach from a circling approach situation, start a climbing turn that will take you back over the airport and then proceed with the published procedure.

The successful completion of a circling approach under really tight weather conditions depends in great measure on your ability to maneuver your airplane in close quarters, and to do it safely. Since radius of turn increases quite rapidly with airspeed, it behooves you to slow down as you enter the final phase of the approach; when the weather is near minimums, it's a good idea to pull back on the reins · a bit anyway so that things don't happen quite so fast. Above all, don't try to bend your airplane around a tight turn to final. Remember that wind direction is probably the reason you're circling in the first place, and as you make the turn to final, the wind will be trying to blow you away from the runway. Steepen the bank and pull back to tighten the turn at this low altitude in less than favorable ceiling and visibility conditions and you just may pull yourself right into the ground. Since stall speed goes up alarmingly with bank angle, you're setting yourself up for an accident; most stall-spin situations that close to the ground provide grist for the headline mills. If you feel uncomfortable while circling, if you get that "gut" feeling that things just aren't what they ought to be, go around and come back for another try.

When you are flying a high-performance airplane, it is possible to get into an unusual situation on a circling approach; suppose your aircraft has a computed approach speed (for approach category purposes) of 89 knots, which allows you to make the approach under Category A minimums. If it is necessary to maneuver at an airspeed that puts you above the 91-knot maximum for this category, you're expected to circle at the MDA prescribed for Category B; it's the circling speed that counts. You probably won't make your low-visibility approach with full flaps (part of "landing configuration"), so adjust the airspeed accordingly.

Here are the points you should remember about circling approaches:

1. Your circling clearance will always specify the landing runway, so you can plan ahead.

2. You must always observe an MDA on a circling approach, no matter what type of approach you started with.

3. You must remain at or above the *circling* MDA for your category until further descent is necessary for landing.

4. While maneuvering, you must keep the airport in sight and remain clear of all clouds.

5. If a missed approach becomes necessary, make a climbing turn that will take you back over the airport; *then* proceed with the published missed approach instructions.

# ☩ The Side-Step Approach

At airports with close-together parallel runways, it's often convenient to have general-aviation traffic use one landing area, heavy jets the other; but it's often just as inconvenient, or operationally impossible, to install a separate ILS for each runway. To alleviate the problem, the "side-step" maneuver was developed. This maneuver provides a legal and safe way to fly the primary approach procedure (which is nearly always an ILS) and, at the last minute, "side-step" to the other runway.

The MDA for such a procedure will always be somewhat higher than the Decision Height for the ILS runway but lower than the circling minimum that would have to be used otherwise.

Controllers seldom issue side-step clearances unless ceiling and visibility are such that you'll have no difficulty finding the landing runway from a mile or so out. If you receive such a clearance and can't spot the runway of intended landing, let the controller know, and your clearance will be amended. Don't try to side-step unless you have the "other" runway clearly in sight.

# ☩ Shortcuts to an Instrument Approach

When ATC gives you the go-ahead for an instrument approach, you are cleared all the way to the missed approach point, and you're guaranteed separation from all other IFR aircraft while in the process of getting there, if controlled airspace exists at the airport. But an approach clearance is not ironbound; it does not mean that you must fly to the MAP in strict accordance with the published procedure.

There are some very efficient ways of shortcutting the involved and time-consuming maneuvers you accomplish flying the full approach, and they're legal, safe, and sensible *when fully understood and properly performed.*

All three of these shortcuts are contingent on breaking out of the clouds at some point well before the MDA or DH is reached. Be familiar with these shortcuts and put them to use when you can save yourself some valuable time and vacate a chunk of airspace for pilots waiting in line behind you. ATC will also be happy because as soon as controllers can get you out of the system they can let someone else in and, in general, speed up the flow of traffic. It's a win-win situation, and everybody is happier. See how far a little professionalism can go?

## "Cancel My IFR"

The first method of shortcutting an approach clearance is to terminate it with a straightforward directive to ATC: "New York Center, Barnburner 1234 Alpha, cancel my IFR." This usually elicits an equally straightforward response from the ground: "Roger, Barnburner 1234 Alpha, your IFR is canceled." Cancellation should not be taken lightly. Be sure you understand that you have now removed yourself from all the protection and services of the IFR system.

Never, never cancel an IFR flight plan and give up all that protection unless you are absolutely sure you can get where you're going in healthy VFR conditions. Controllers look with understandable scorn on pilots who cancel IFR and then return sheepishly for a clearance after finding out that they can't maintain VFR. Although the people in the Center are there to help, they don't like to duplicate their efforts, and such a request is sometimes answered with, "Call the nearest Flight Service Station and file a flight plan." Now you must go back through the system, maintaining VFR all the while, and it's going to cost you minutes on the clock, gallons on the fuel gauge, and maybe a missed appointment. Anyway you look at it, an unwise IFR cancellation is spelled M-O-N-E-Y.

You must be especially cautious about cancelling after the sun goes down. The next pilot who is lulled into a false sense of "I've got it made" when the rotating beacon at Destination Municipal becomes visible won't be the first. The pilot cancels, maybe even with the airport itself in sight from afar. Nothing is so reassuring as those flashes of white and green beckoning to you out of the darkness after a long trip, and the feeling of security grows as you pass over the airport, able to see the lights, the windsock, and all the familiar landmarks. Now onto the final approach, and just about the time you flare, the whole world turns white. Visibility was unrestricted when you were looking straight down through the fog, lying in gossamer sheets just above the runway, but when you flatten out to land, you're looking through the fog layer endwise and the visibility goes to zero. Not only is this situation embarrassing if you have to return to ATC for the clearance you gave up so confidently a while ago, but there is also an instinctive reaction to push the nose over when encountering suddenly lowering visibility. When you're that close to the ground, it may be disastrous.

## "How About a Contact Approach?"

The second shortcut is very much like a Special VFR clearance, using the same basic weather and navigation criteria. Known as a "contact approach," it is unique in that it must be requested by you, the pilot, when you feel you can navigate to the airport by ground references, thereby eliminating the need to fly the complete published procedure. It's a real timesaver, and whenever you can lop a few minutes off the time you spend in the air, you're using the airplane more efficiently and opening up the airspace for other pilots in the bargain.

Since it is a deviation from the instructions you have received, you *must* receive a clearance for a contact approach. The rest of the way to the airport, it's up to you to navigate, to look out for other aircraft, and to stay at a safe (and legal) altitude. ATC will not turn you loose unless you are safely separated from other aircraft on contact approaches, regular instrument approaches, or those operating under Special VFR clearances. The Tower will probably clear you to land on the first communication, and you have just saved yourself the trouble of going to the navaid and flying the complete published procedure, which may have taken you another 5 to 10 minutes.

The specifications for a contact approach call for you to be reasonably certain that you can get to the airport on your own. But what happens if you are unable to maintain the 1-mile visibility/clear of clouds restriction? You must let the controller know, and you'll be advised to climb to the appropriate minimum altitude and continue with the published approach procedure. You are still under the protective umbrella of the IFR system and may continue just as if you had never requested the contact approach in the first place.

In summary, then, the conditions for this shortcut are:

1. The *pilot* must request a contact approach.
2. ATC must issue a specific clearance, that is, "cleared for a contact approach."
3. You must be able to maintain at least 1-mile visibility and remain clear of all clouds.
4. You must be able to navigate to the airport by means of ground references.

As always, you are expected to avoid other aircraft whenever you are able to see them.

## The Visual Approach

The third deviation from a published instrument approach procedure is the visual approach. Although it accomplishes the same thing as the other shortcuts, there are significant differences. In the first place, a visual approach is almost always instigated by ATC (although there's nothing wrong with requesting one if it will speed things up). You will be radar vectored to a point at which one of two things will happen; you'll see the airport or you will be directed to follow another airplane, and the controller will ask you to verify that you have the other airplane or the airport in sight. More closely controlled than a contact approach, the visual approach is intended to lead you directly to the destination airport, and it shortcuts the full approach at the discretion of ATC. The controller must know that you are in VFR conditions before issuing such a clearance; that's why you're asked if you can see the airport or the airplane ahead. When it's honest-to-goodness IFR, plan on the complete approach.

On occasion, a visual approach clearance is issued when you are behind another IFR flight approaching the same field. When the controller sees that you are reasonably close to that airplane, you'll hear something like, "34 Alpha, do you have the Trans-Lunar 727 in sight, 1 o'clock and 3 miles?" When you answer in the affirmative, you'll be directed to "follow the 727, cleared for a visual approach," and your responsibility is to do just that, hoping all the while that the Trans-Lunar pilot is headed for the right airport.

With no charted instructions, each visual approach is unique in terms of track and altitude. You are expected to continue toward the airport, and if the destination is nontowered, you may be cleared straight-in or directed to enter a normal traffic pattern. In the case of a visual approach to a nontowered airport, remember that you have become just another participant in VFR operations at that airport. The use of standard VFR practices, position announcements, and see-and-avoid tactics is mandatory.

Most likely, you'll receive clearance for a visual approach while flying at an assigned altitude. When the controller clears you for the visual approach, altitude management is completely in your hands, and you should descend or stay level in accordance with your assessment of weather conditions, distance from the airport, terrain, obstacles, other traffic, and so forth.

The visual approach presents ample opportunity to cheat because you might answer, "Roger, I have the airport [or the Trans-Lunar 727] in sight" when you don't, but you're just cheating yourself because you're in VFR conditions, and ATC is expecting you to provide your own separation. There may be a considerable number of VFR pilots milling around in the 3-mile murk (in busy metropolitan areas, you can *bet* on it). You are a definite hazard if you accept a visual approach clearance when you're not really "visual."

Here are the conditions for a visual approach clearance:

1. Usually instigated by ATC.
2. Weather minimums, basic VFR.
3. Will not be issued until you have the airport or a preceding aircraft in sight.
4. Puts the "see-and-avoid" monkey on *your* back when the clearance is accepted.

Controllers will never ask you to cancel your IFR flight plan. After all, handling instrument traffic is their job, and they are obliged to provide all the services of the system to any pilot who operates IFR. But they sometimes make a request like this: "Barnburner 1234 Alpha, will you accept a contact [or visual] approach?" It's the controller's way of trying to tell you that a good purpose can be served by having you fly a shortcut to the full procedure. Never construe this as a goad, but make certain that you are in a position to comply with all the requirements of the situation before accepting. If this is the flight on which you promised yourself you would practice a full ILS right down to the deck, by all means tell ATC that you would prefer the complete approach. Common sense and courtesy suggest that you don't try this at LAX at 9 o'clock on Monday morning; although your request for a full approach will probably be granted, you might hear, "Roger, 1234 Alpha, cleared direct to the outer marker, climb to and maintain 8000, hold east on the localizer course, expect approach clearance sometime tomorrow."

The hazards of a visual approach are underscored rather dramatically by the sequence of events one day at a southern Michigan airport. A Skymaster, inbound from the southeast with the field in sight, requested and received a visual approach clearance when still 7 miles from the airport. At about the same time, a Skylane departed for home, which lay some 30 miles southeast. The tower controller did not have the benefit of radar to help locate and separate flights in the airport traffic area.

Both pilots were advised of the other's location, heading, and intent, with no acknowledgment of sighting by either. The Skymaster pilot was a middle-experience commercial pilot, flying alone; in the right seat of the Skylane was a 4000-hour CFII, working with an IFR student, who was probably under a hood.

Neither of these aircraft presents much of a head-on profile, and almost like moths drawn to a flame, they continued toward each other, one climbing, one descending, until at the very last moment, each pilot appeared to spot the other airplane. The Skylane rolled violently to its right (according to eyewitnesses), and the Skymaster rolled to its left, making a bad situation worse. There was almost time for the Skymaster pilot to realize what had happened and roll back to the right, out of trouble. But "almost" wasn't enough; the Cessnas tangled left wings, and both fell to the ground, out of control.

Within the limits of the rules under which they were operating, these pilots could have taken steps to reduce the possibility of a collision; the Skylane pilot could have stopped his climb until clear of the traffic, and the Skymaster pilot could have leveled off until clear. Either pilot could have changed heading to right or left (with appropriate advice to the controllers) to resolve the head-on situation. .

Is such an encounter a potential hazard of nearly *every* visual approach? The answer is an unequivocal "yes." The very weather conditions that spawn requests for and issuances of visual approaches (that is, anything between 1000 and 3 and "severe clear") place the instrument pilot right in the middle of normal everyday VFR operations. There's no room for complacency here. Even though you have received an "instrument approach clearance" and are operating under the Instrument Flight Rules, you are just another VFR flight when it comes to "see and avoid." *Whenever* you are able to see other aircraft, and therefore avoid them, you are expected to do just that.

# ✛ Radar Approaches

"1234 Alpha, this is Disneyland Approach Control, turn right heading two two zero, vectors for a surveillance approach to Runway 18." Should you hear such a transmission in today's IFR world, chances are the Barnburner is either doing practice approaches or is in trouble; the use of radar for actual approaches has dropped off to just about zero. Radar "did itself in" by virtue of the very feature that promoted it many years ago. The fact that an aircraft could be brought safely to Earth if the pilot had basic instrument skills and one communications receiver led some aviators to practice self-conducted approaches only for emergencies, such as a radar failure. But it was not in the cards because air traffic has increased so much that communications channels would be totally clogged if every IFR flight had to be handled one on one. A radar approach requires a lengthy and nearly continuous stream of instructions and corrections from the controller, limiting system capacity to one airplane at a time; air traffic would be backed up all the way to Constantinople if all the flights coming into JFK had to be radar controlled.

So the tables have turned on radar. Once hailed as the best way to get out of the clouds and onto a runway, it now takes a distant back seat to the cockpit-display type of approach in which one controller can point several pilots in the right direction and turn them loose, trusting to their navigational skill and keeping communications channels open for directing other flights. But like an understudy waiting in the wings for the leading lady to falter, radar is always standing by for that situation when somebody *needs* a bit of vocal assistance.

Airport Surveillance Radar (ASR) is primarily an approach control system. ASR beams sweep out a considerable distance and depict the location of air traffic throughout the terminal area. Coincidentally, it can also be used as a means of vectoring aircraft to the final approach course for all runways at the airport. (Some installations are limited to certain runways because of terrain features, electronic problems, and the like.) ASR displays azimuth, distance, and altitude, but with no glide slope information, it provides a nonprecision procedure. Therefore, an ASR approach will be flown to a Minimum Descent Altitude (MDA), not a Decision Height (DH).

When you're cleared for an ASR approach, you'll be given headings to maintain the final approach course, but altitude, airspeed, and rate of descent are up to you. On request, the controller will compute a recommended rate of descent for you and may provide recommended altitudes as you fly toward the runway. But just like any nonprecision approach, it is best to get down to the minimum altitude as soon as practical; it gives you more time to look for the runway and prevents your arriving over the airfield in solid clouds 50 feet above the ceiling, requiring a needless missed approach. Most ASR approach procedures specify a visibility minimum of 1 mile; when the runway environment includes certain types of approach lighting systems, that minimum may be reduced to as little as one-half mile.

An ASR approach is basic instruments all the way and, as such, makes a great introduction to approach work for an IFR beginner. Most terminals equipped with approach control radar will also be set up for surveillance approaches, and the controllers will be happy to accommodate your practice whenever traffic permits.

When weather is not a factor and traffic is light, controllers will usually grant your request for a radar approach. They realize that it is a good training exercise; as a matter of fact, there are times when the request comes from ATC because a certain number of approaches must be conducted to keep the controllers qualified. You have an interest in the skill of the people on the other side of the radar scope (Don't you want them to be *good* when it's the only way to get out of the sky?), so go along with such a request whenever you can. There's not much to be gained by practicing radar approaches frequently; you will have proved that you can do very precisely what someone else tells you to, but your instrument training time is better spent on other types of approaches, the ones where all the navigational responsibility is on *your* shoulders. So, know what a radar approach is all about, do one for practice now and then, but consider it something you will do for real only when everything else has turned sour.

# ✝ The No-Gyro Approach

There was a time when instrument pilots had to prove that they could turn accurately from one heading to another using only the magnetic compass. And one of the biggest reasons for that demonstration rested in the directional gyros (DGs) of the day, prone to precessing wildly or failing altogether, especially those powered by venturi systems, the predecessors of today's vacuum pumps.

On occasion however, even a modern vacuum pump gives up, and there you are—in the soup—trying to maintain a heading to follow an approach procedure but finding it impossible because none of the numbers make sense. Radar to the rescue again, this time with a procedure known as the "no-gyro approach," in which the controller will issue clear-cut instructions to "turn right," "turn left," and "stop turn." You are expected to execute standard-rate turns on command because the controller is timing your turns to result in headings compatible with the approach procedure.

If the heading indicator is not working, chances are very good that the attitude indicator is equally unreliable (most general aviation aircraft are provided with a common source of power for these two instruments). So you might as well shift gears and fly partial panel for the remainder of the approach (see Chap. 4, Partial Panel IFR). When you're lined up on the final approach course, the controller may request half-standard-rate turns; you're proceeding into the neck of the approach funnel, and heading corrections must be made on a smaller scale.

A no-gyro approach represents a great opportunity for overcontrolling. You must settle down and fly the airplane with the finest touch you can muster. It's needle-ball-and-airspeed the rest of the way—and it's also too late to practice.

## ✝ "Below Minimums" Approaches

Military and airline pilots are prohibited by regulation from even *starting* an instrument approach when the weather at the airport is reported below their landing minimums. A different approach to the problem shows up in Part 91, not-for-hire operations, where the only limiting factor is visibility; if you fly down to the appropriate altitude (MDA or DH) and can see the runway environment, you are within your legal rights to continue the approach to a landing. This philosophy reflects sound thinking on the part of the regulation writers, since it allows noncommercial pilots to make their own inflight visibility observations. And who is in a better position to decide? As long as you respect the published minimum altitude for the approach and gain sight of the runway at the appropriate time, the ceiling requirement will take care of itself.

Though making an attempt (the so-called "look-see" privilege of not-for-hire operations) is not always the smartest thing to do, a pilot operating in accordance with Part 91 may not be denied an approach clearance solely because the weather is bad.

## ✝ It's All Up to You

Here is where an inflight visibility determination can really work in your favor. Suppose the reported visibility is three-quarters of a mile, and you have eased down to the MDA well ahead of the missed approach point. When the clock tells you that you're a mile out (or the appropriate distance on the chart), you look up and see the runway lights shimmering through the fog. Is it legal to land? It certainly is, as long as you can keep the runway lights in sight the rest of the way. The airport layout often furnishes a clue to the disparity; control towers are seldom located near the approach end of the runway, and the controllers' visibility may indeed be only three-quarters of a mile, which prevents them from being able to judge conditions on the final approach course. You are in a *much* better position to decide.

There is a decided note of permissiveness in this clearance, and as such, there is a very big loophole: You can press on in reported lower-than-minimum conditions, and if you do a good job of navigating, chances are at least fair that you will eventually see the airport and be able to land. But don't forget that everything is up against the limits, and you've got to do it all right the first time. For one thing, when the visibility is such that you don't see the runway until you're right on top of it, the distance you'll need to touch down and get stopped is automatically increased. Restricted visibility is often caused by some type of precipitation, which means that the runway will probably be wet; at worst, it may be covered with ice or snow. And every second you fly, the required rate of descent increases. The wind will invariably choose this instant to die, with the result that the distance you need for landing may exceed the runway available.

So back off and take a good long look at all the factors before you commit yourself in conditions like these. If your first approach finds you too high, too hot, and too hazardous, execute the missed approach procedure and come back at a slower speed, mentally prepared to pick up the chips. You can't be faulted if you abide by the rules, which give you the benefit of dynamic observation.

We'll let the NTSB report set the stage for this one:

"According to witnesses, the aircraft touched down about two-thirds down the length of Runway 04. Skid marks began approximately 1,000 feet from the end of the runway. These skid marks tracked the runway centerline until about the last 300 feet, when the marks began to drift slightly to the right of centerline. The aircraft then skidded off the runway and continued approximately 100 feet through a grassy overrun where the nose landing gear was sheared off. At this point the ground drops off abruptly. The aircraft, now airborne, sailed another 130 feet before striking the ground in a near-vertical attitude...flaps were deployed one notch, a setting normally used for short-field and soft-field takeoffs."

This accident (which killed the pilot and seriously injured his passenger) didn't involve an instrument pilot, but it illustrates the folly of a last-minute decision when a missed approach was clearly indicated.

The pilot approached the airport in VFR conditions, but rain showers in the terminal area obscured the field until the airplane was just one-half mile away. The controller had suggested 09 Right as the landing runway, but when the airplane came into view so close, he transmitted, "Okay, if you can make Runway 04 from that point, that's fine. If not, just keep circling and make left traffic for that runway." As previously described, the pilot touched down on Runway 04, but with a 20-knot tailwind, partial flaps, and higher-than-normal approach speed (as judged by a witness), he didn't stand a chance.

There's a lesson here for IFR pilots because this is the very situation that shows up not infrequently at the end of a circling approach to a strange airport. A perceived need to get on the ground on the first attempt, strong wind (which probably dictated the circling approach), and an unwillingness to confess and climb away for another try have resulted in embarrassment for many pilots; for this one, the cost was considerably greater.

# ✝ Speed Adjustments During an Approach

The first priority of ATC controllers in a terminal area is to provide safe separation for aircraft arriving and departing, and as the number of flying machines grows, the control problems increase accordingly. With more operations in the system everyday, controllers find themselves involved in valiant efforts to sort out the speed requirements and performance limitations of a mixed bag of aircraft. At a busy airport, approach controllers are faced with funneling everything from 747s to 152s into the narrow confines of the approach corridor; speed and timing is the name of the game. A veteran railroader visiting an approach control facility noted the multitude of aircraft approaching from all directions at all speeds, flying into the "small end of the funnel" onto a single runway. He could not help comparing this operation with his own experience, in which the main line spread out into many divergent rails in the terminal, allowing trains to come to a stop on any one of a dozen or more tracks. He allowed that trying to put all those airplanes coming from every direction onto one runway was "a hell of a way to run a railroad!"

It isn't likely that we will ever see multiple-runway airports, so we're stuck with adjusting speeds and timing to fit air traffic into the approach funnel. Controllers will cut and fit to achieve their objective and are furnished a set of guidelines with which to do the job. You needn't know what these guidelines are, only that they exist, so that a request for a speed adjustment won't come as a complete surprise. The jets and some of the faster propeller-driven aircraft have little concern with minimum airspeeds and are more often than not asked to *reduce* speed, but the light airplane fleet is sometimes required to add a few knots in order to fit into the schedule of approaches; that "few knots" sometimes means maintaining near-cruise airspeeds to the Final Approach Fix. Controllers are aware of the limitations of light aircraft and would not likely ask a Musketeer pilot to maintain 160 knots, but in most of the light twins and some of the high-performance singles, these speeds are not at all unreasonable. In any case, you should deal with speed adjustments as *requests*; if a controller asks you to fly at 150 knots and there's no way the you can do it practically or safely, reject the request... politely, of course.

# ✛ The Stabilized Approach

A pilot's first experience at the controls of a jet airplane is usually an exercise in humility. The tremendous increase in the speed at which things take place usually results in the pilot getting "behind the airplane." For example, when the jet is climbing through 500 feet, the pilot is still on the runway, at least mentally.

Being behind an airplane is a lot like flying an unstabilized approach, although the latter is usually the result of bad planning, lack of proficiency, or a burning desire to get on the ground in a hurry. Either way, an unstabilized approach makes a normal procedure very difficult, and a pilot's attempt to salvage the situation sometimes produces an incident at best, an accident at worst. A bad start often results in a bad finish.

Before we go further, here's a general definition: A stabilized approach is one in which the airplane is properly configured in all respects when it reaches the point where descent to the minimum altitude begins, and one in which there are no major deviations from course, altitude, or airspeed throughout the approach. Configured "in all respects" means on course, at the proper altitude and airspeed, with landing gear and wing flaps as required. When you reach the descent point, you can start downhill by reducing power, lowering the wheels, or both. Everything is solidly under control, and there's no need to chase the needles, jockey the throttles, or cope with large trim changes when you should be concentrating on precise and accurate instrument flight. Here are some techniques to help you fly a stabilized approach.

First, plan to be established at the appropriate minimum altitude at least 3 miles prior to the Final Approach Fix or glide slope intercept. If you're new at instrument flying or the weather conditions are exceptionally bad, you may want to increase that distance to 5 miles, giving you more time to get stabilized. If you're too high or too fast or both, tell the controller that you need a 360 to lose altitude or airspeed. Better yet, try to recognize this problem early on and conduct your descent accordingly. On a busy day, you may be sent to the back of the line if you request a delaying turn.

Second, when you reach the 3-mile point, get the airplane configured for the descent to minimum altitude. In fixed-gear airplanes, most pilots don't bother with wing flaps on an IFR approach, so your primary consideration should be airspeed (if you'd rather use a partial flap setting, set it up now; it's one less trim change you'll need to make later on). In a retractable-gear airplane, there are two schools of thought: You can lower the gear now and let the airspeed stabilize or you can wait until the final approach fix or glide slope intercept and lower the wheels to begin the descent. It's a matter of personal preference, so take your choice.

Third, at the FAF or glide slope intercept, power back or lower the gear to start the descent, and from here on, you need make only small changes in power and heading to keep everything where it should be. The stabilized approach provides a smooth transition from level flight to controlled descent, and your chances of getting way off course or glide path are greatly reduced. When you fly stabilized approaches, your only missed approaches will be those caused by truly lousy weather.

A commuter pilot (and his passengers) learned about stabilized approaches the hard way. Relatively inexperienced in the airplane he was flying, and known to be a regular user of the autopilot because of his lack of confidence in his own IFR skills, this pilot arrived at the glide slope intercept with much more airspeed than the manual called for, and well above the maximum gear operating speed.

When the autopilot sensed glide slope intercept and pitched the nose down, the pilot brought the power levers all the way back to slow down so he could lower the gear. That worked, but it worked too well; the drag of two big turboprop engines at flight idle caused the airspeed to drop off at an alarming rate, and the pilot was apparently so distracted by the airspeed problem that he forgot to power up again.

By this time, the pilot had also extended partial flaps, and when the airplane slowed to the point that the autopilot could no longer hold it on the glide slope, it disconnected. According to the comments on the cockpit voice recorder, neither pilot realized what had happened. When the stick shaker and then the stick pusher activated, the pilot knew he was in deep trouble, so he pushed the power levers forward, hauled back on the yoke and, incredibly, called for "flaps up." As you might suspect, the airplane stalled and fell out of the sky, crashed into a building a mile or so from the runway and burned. Both pilots and several passengers died.

How will you know when a stabilized approach is going sour? They do sometimes, you know. When the needles—localizer, glide slope, VOR, whatever—move away from center and stay there or, worse, keep on going, you're flirting with destabilization, and you should seriously consider going around. This is dependent, of course, on how far and how fast the needles move. The same is true of airspeed, altitude, and rate of descent; when any one of these starts to get out of control, power up and go around.

Perhaps the best indicator of an impending problem is your personal feeling about the progress of the approach. If it just doesn't feel right, just doesn't look right, it's time to break it off and try again. The next time around, you'll know what to expect, and you'll probably do a much better job.

# 16

# The NDB (ADF) Approach

An ADF receiver, adequate training and practice, and a couple of grains of common sense will outfit you splendidly for an NDB (NonDirectional Beacon) approach. (Everybody knows what you're talking about when you refer to it as an ADF approach, but the official title is NDB, and rightly so, for it's the radio receiver in your airplane that is the ADF, or Automatic Direction Finder.) This procedure enables you to fly to a point on a specific course at a prescribed altitude and from this point continue to a landing if visibility is sufficient. It's not the most accurate method of finding an airport because it requires considerable skill and technique on the part of the pilot, but it's a lot better than no approach at all. Even with this lack of precision, a typical NDB approach will bring you to within 500 or 600 feet of the airport surface, and the visibility required for landing will be about 1 to 1.5 miles.

At the heart of all ADF procedures is "relative bearing," a term that could no doubt stand a bit of amplification. The azimuth indicator (needle) of the ADF set will always point to a number on the instrument that represents the bearing from aircraft to station, measured relative to the nose of the airplane (0 on the ADF dial); hence the term relative bearing. When the relative bearing is O, the station is straight ahead; should the relative bearing turn out to be 90 degrees, the station is off the right wing tip; at 180 degrees, you'd have to turn around in your seat to see the transmitter; and a relative bearing of 270 degrees puts the station off the left wing.

With this constant readout of relative bearing, the navigational task becomes one of determining whether you are on, to the left of, or to the right of a preselected course to or from the station. When you are on course, flying toward the station with no wind to push you off course, the aircraft heading will be the same as the desired course, and the ADF will read 0 degrees relative bearing. Pass directly over the transmitter, and the needle will immediately swing to the 180-degree, or tail, position; if you really want to make it easy for yourself, think of it as a 0 degree relative bearing *outbound.*

A touch of plane geometry will help you understand what happens when you intercept a course with ADF, as well as the mechanics of drift correction. Some geometrician discovered early in the research into the properties of straight lines that, whenever two of them are crossed, the opposite angles are always equal. You can put this knowledge to work on an ADF problem by thinking of the intended course to the station as one straight line, with the aircraft's heading as the other. As long as the lines are superimposed (on course, no wind), there are no angles and no problem, but if the aircraft heading changes, the ADF needle will move, creating an angle between heading and track.

Conjure up this situation: You are on course, headed to the station, relative bearing 0 degrees. Now turn the airplane 30 degrees to the left; the ADF needle, always pointing to the station, will appear to move to the *right,* exactly 30 degrees; the opposite angles are equal. So, whenever you are *off* course and are flying at an angle to that course so as to intercept it, you'll know you are *on* course when the angles are the same. By using the same principle, you can remain on course with a drift correction that will keep the two angles equal. Using the intended track as the base, you're *on course* whenever the heading and relative bearing are displaced the same number of degrees. *Inbound,* heading displacement will be *opposite* that of the ADF pointer; *outbound,* heading and ADF needle will be displaced on the *same side.*

## ✝ Tuning the ADF Receiver

Radio stations upon which ADF procedures are based are generally called NDBs, for NonDirectional Beacons, which suggests that the signals from these transmitters may be received anywhere around the antenna; the signal is not concentrated, or aimed, in any particular direction. The NDB frequency range is 190 to 415 kilohertz; the receiver can also be used as an AM radio in case you want to listen to music, news bulletins, or ball games.

The function switch should be in the REC (RECeive) position whenever the set is being tuned. It may be marked ANT (ANTenna) on some radios, but it means the same thing; in either case, the Automatic Direction-Finding function is not operating until you switch to ADF. Tuning in the ADF position means that the needle will try to point to every station it receives as you turn the dial. Just like a bird dog trying to point every quail in a covey, it doesn't give you much useful information, and it's hard on the dog! Give your equipment a break by finding the proper frequency on REC or ANT, then switch to ADF, and let the needle do its thing.

With the exception of the few remaining high-powered beacons, don't expect a usable signal until you are within 15 or 20 miles of the station. Most NDBs are intended only for approaches and are not powered for long-range reception, so don't complain to ATC about the quality of the transmitter because you can't pick it up 50 miles out. The controller will somehow get you within range before turning you loose to navigate to the beacon.

There is only one way to make certain you have tuned the right station, and that's by listening to the Morse code identifier. A great many ADF stations are crammed into a rather narrow band of frequencies, so it's easy to get a needle indication that looks okay, but it's mighty embarrassing to make an approach on the wrong station. It could also cause you to aluminum-plate a hillside or reduce the height of a radio tower or two.

# ✝ Initial Approach

Getting to the NDB to commence the approach is a piece of cake with radar vectors, but you'll have to put on your navigator's hat when the controller clears you "from your present position direct to the beacon." Since the ADF needle always points to the station, you can turn the appropriate number of degrees to put the needle on the nose (or 0 degrees relative bearing), keep it there, and you'll wind up over the station. True, and on a windless day you'll fly directly to the station, which is just what ATC wants you to do. But throw in a crosswind, and the picture changes. Now instead of making good a direct line, or track, keeping the needle on the nose will result in a curved path across the ground. This is called "homing," is done best by a specific breed of trained bird, and is not at all what is expected of you; you were cleared *direct,* and ATC would like you to go in a straight line. Leave homing to the pigeons.

Here's a typical situation. You are cleared "direct to the beacon," your heading is 360 degrees, and the ADF needle is 40 degrees to the left of the nose. A left turn of 40 degrees will put the needle on 0 and the station directly ahead of you, and that's fine, as long as it lasts. But a right crosswind will eventually drift the airplane off course to the left; you maintain a heading of 320 degrees, and the needle shows the drift by moving slowly to the right. Here is a key point: Whenever the aircraft heading is the same as the desired course, and the ADF needle is off to either side, *it always tells you which way to turn to get back on track*. In this case, it indicates that your intended *track* of 320 degrees is to the *right*. The ADF needle has become a *course pointer*.

You must obviously set up a drift correction, but first get back on course by turning 30 degrees to the right. When this new heading puts you on the intended course, the ADF needle (always pointing to the station) will point exactly 30 degrees to the *left* of the nose (opposite angles equal). Had you turned right 45 degrees to get on course faster, the ADF will indicate 45 degrees to the left when you are on course. The number of degrees turned is immaterial as long as you remember that a corresponding number of degrees to the *opposite* side on the ADF dial will tell you when you're on course again.

Now apply some drift correction. Start with 10 degrees. As long as your heading is 330 degrees (the intended course of 320 degrees plus 10 degrees drift correction) and the ADF reads 10 degrees left, you're on course, and the drift correction is just right. By using the cut-and-try method, you'll soon find a heading that will balance drift correction and wind effect, and you are proceeding "direct" to the station, as cleared. (You should apply the principle of course bracketing; see Chap. 9.)

Drift correction is not so difficult if you will think of your situation as one in which heading and ADF needle displacement must remain balanced. If there's no wind, your heading will be the same as the desired course, and the ADF needle will read 0; no angle between the two straight lines, everything in balance. When a 15-degree drift correction to the *left* is needed to stay on track, it will be reflected in your heading, and to balance that, your ADF should read 15 degrees to the *right*; opposite angles equal, everything in balance. A change in ADF indication (maintaining a constant heading, of course) means that your drift correction is not doing the job; if it creeps back toward the nose, you haven't enough crab, and if it goes the other way, you have overcorrected.

A directional gyro that includes some sort of heading "bug" is really a blessing because it helps you solve the memory problem. As soon as your track to the station is determined, set the bug on that number, and you can tell at a glance how many degrees you have turned to counter the crosswind. Another glance at the ADF indicator will confirm the balanced situation or show the need for another heading change.

# ✝ Station Passage and Tracking Outbound

When the ADF needle swings past the wing-tip position on either side, you have passed the station. Assuming for now that a full approach procedure is required, you must track outbound on a specified course, turn around, and come back to the airport. There are a number of things to do in addition to flying the airplane, so get yourself organized.

As soon as you recognize station passage, use the four Ts: *Time, Turn, Throttle, Talk*. Note the *time* (a timer that shows elapsed time is a big help), *turn* to the outbound heading (you can make drift corrections in just a minute; the outbound *heading* is more important now), *throttle* back to begin the descent to procedure turn altitude (if you're not already there), and finally, after everything has settled down and is under control, *talk*—tell the controller you have passed the beacon outbound. You can make it a very brief report: "Tampa Approach, Barnburner 1234 Alpha, KAPOK outbound."

Now, if drift correction is required, put it in; remember that when tracking *outbound* the ADF needle will point rearward, and the "balance" between heading and ADF indication will be on the *same side*. If the outbound course is not significantly different than your inbound track to the station and if the wind stays the same, your 10-degree drift correction will now show up on the ADF as a needle indication 10 degrees to the *right* of the tail. In other words, when the outbound *heading* is off 10 degrees to the right to correct for wind and the ADF needle is *also* off 10 degrees to the right, you are on course. If things stay just like that, you have the proper drift correction.

To intercept a track outbound, turn to the new heading and note the position of the ADF needle relative to the *tail*. If the needle is off to the right, that's where your new track lies, and vice versa; so head in that direction, with one minor change: When outbound, it's important that you get on course as soon as possible, so turn 45 degrees toward the needle right off the bat. When the ADF shows a 45-degree deflection on the *same side,* you are on track; return to the desired heading, plus drift correction if needed. Try 10 degrees, and as long as the ADF needle stays 10 degrees to the same side, everything's ship-shape. If it creeps toward the tail, you've too much wind correction; if it moves away from the tail, too little. Even though your ADF dial is imprinted with "18" at the tail of the aircraft, think of it as just another "zero point," and you'll be able to handle the problem more easily.

# ✛ Procedure Turn

When it's time (2 minutes outbound is a safe, practical figure in most cases), begin your procedure turn in the direction published on the chart; the headings outbound and inbound are always printed, so turn directly to the proper heading. If you started the turn on course, the ADF needle will have moved 45 degrees from the tail when you roll out on the published heading (opposite angles equal) and once again note the time. If you miss the 45-degree indication, start the clock when you level the wings, and you'll not be far off. (By the way, have you noticed the index marks at the four 45-degree points on the face of your ADF dial? They were put there for "standard" situations like this. If your ADF dial is indexless, there's no law against painting a set yourself.)

Fly outbound for 1 minute (or 40 seconds plus or minus wind correction if you prefer that type of procedure turn) and then turn 180 degrees in the opposite direction to intercept the final approach course. If ATC has requested that you report "procedure turn inbound," now is the time to do it. When you are on course again, the ADF needle will be 45 degrees from the nose, and it's back to a simple problem of tracking to the station. If appropriate, you may now descend below procedure turn altitude on your way to the next altitude published for the approach.

# ✛ Postprocedure Turn
# and Final Approach

You've a couple of minutes to kill inbound to the station, but don't get behind the airplane. There are four things you can do here: First, compensate for wind drift; most final approach courses (from station to airport) are the same as the inbound course you are on at this point, and wind direction *generally* becomes more southerly as you descend, so anticipate it. Second, slow the airplane to approach speed. Third, configure the airplane for the final approach—gear down, flaps as required—so all you'll have to do at station passage is reduce power to descend comfortably to minimum altitude. Fourth, recheck the DG against the magnetic compass reading (be certain you do it with wings level and airspeed constant). You can fly a beautiful NDB approach, with all the numbers appearing correct, and then break out of the clouds looking down the wrong runway or, worse, seeing no runway at all. Remember that unlike VOR navigation, in which a centered needle means you're on the correct magnetic course regardless of heading, an NDB approach is only as good as the heading information on which it's based. If you're lucky, an approach on the wrong course will merely surprise you; if you're *un*lucky, well, nobody wants to get that intimately acquainted with a mountaintop or a TV tower.

When the needle swings, note the *time* (the missed approach point is always based on timing on an ADF approach, unless the beacon is located on the airport), *turn* to the final approach course if necessary, *throttle* back for descent, and when a comfortable rate of descent is established, *talk,* "Tower, Barnburner 1234 Alpha, KAPOK inbound." Nine times out of ten, you'll be cleared for landing.

Now you're down to the nitty-gritty of the approach, and your objective is to get to the MDA as soon as practical, so that you can look for the ground. Be sure to arrive at MDA well ahead of the missed approach point; it's embarrassing and inefficient to miss an approach because you didn't get down to the MDA soon enough.

Your rate of descent from final approach fix to MDA is based on the estimated groundspeed for this segment, and the surface winds from the Tower or AWOS are as good an input as any at this point. Your approach chart furnishes the time-distance figures (Jeppesen provides the required rates of descent for various groundspeeds), but it's a better practice to get down to the MDA well in advance of the missed approach point. You are guaranteed terrain and obstacle clearance for the entire final approach segment if you remain at the MDA. The sooner that you can see the ground, the better off you will be, so set up a comfortable rate of descent (certainly not more than 1000 feet per minute) that will get you to the MDA well before the missed approach point. When you get there, *stay there* until either you can see the airport or your time has elapsed. Airport in sight? Go ahead and land in accordance with your clearance or the recommended VFR procedures if you're approaching a nontowered airfield.

But suppose the clock hands race around to the appointed mark, you're at MDA, and still "in the suds." *Don't* hesitate, *don't* descend another 50 feet to take a look, *do* execute the missed approach procedure *now!*

The final segment of any approach procedure is the most critical in terms of your airplane's relation to the desired course; you need to be *on course* if the procedure is to accomplish its objective. And perhaps more than approaches based on other types of radio aids, the NDB approach is loaded with potential for confusion and wrong-way corrections. Until you work out a foolproof method of analyzing the navigation indications and staying on course with no doubt about your location, don't be the least bit hesitant to make an occasional turn back to the published course to check the ADF needle deflection.

If the airplane is *on course* when you accomplish a "course pointer" check, the ADF needle will point to 0 or 180 (inbound or outbound tracking, respectively) as soon as your heading is the same as the published course, and you should immediately return to the former heading, since whatever drift correction you had established is evidently doing the job.

If the airplane has drifted off course, a quick check will show you which way to correct; the head of the ADF needle will *always* point toward the desired course when that course is paralleled.

A couple of these checks during the final segment beats the socks off no checks at all or correcting the wrong way and making a bad situation worse. You'll feel much better about your ADF work when you have the confidence that comes with sure knowledge of your position.

# ✝ The Radio Magnetic Indicator (RMI)

NDB approaches are like shooting fish in a barrel if you are fortunate enough to have a Radio Magnetic Indicator installed in your airplane. The RMI is a slaved gyro heading card that moves under the ADF needle. (Some installations include a second needle that points to a VOR.) Now, instead of the non-RMI presentation in which 0, or the nose of the airplane, is always at the top of the dial, the aircraft's magnetic *heading* appears under the top index; therefore, the ADF needle *always* points to a number that represents the magnetic course to the station.

Referring to the previous example, when you are cleared direct to the beacon, the ADF needle would still indicate that the station is

40 degrees to the left of the nose, but it would also show that the course to the station is 320 degrees. When you turn to the left, roll out on that heading (320 degrees), and as long as the needle stays on 320 (which will also indicate a zero relative bearing) there is no drift, and you are proceeding directly to the NDB. If wind begins to drift you off course, the ADF needle will again tell you which way to turn to get back on track. But with the RMI at work for you, the correction is much easier. Use the same method of turning toward the needle, but rather than having to go through the mental calculation of balancing heading change and relative bearing, hold your into-the-wind correction until the needle once again reads 320 degrees. From here on, do whatever is necessary to keep the RMI needle on 320 degrees, and you can rest assured that you're tracking directly to the station. (You may notice that whenever the ADF needle is on the prescribed course, it will also show a relative bearing that balances the drift correction, but you don't have to figure it out because the RMI does it for you.) Track outbound the same way, keeping the needle over the number that represents course to the station (it will be the reciprocal of your inbound track), and you will be on course. The inbound portion of the procedure turn is made much easier with an RMI; just fly on the inbound heading until the needle points to the inbound course, at which time you turn to that heading and keep the needle where it belongs.

# 17

# The VOR Approach

By the very nature of the electronic equipment involved, the VOR approach climbs a rung higher on the accuracy ladder, but it still must be ranked with the nonprecision procedures because it lacks a glide slope.

The number of VOR approaches has grown steadily through the years as the criteria for an instrument procedure are met at more and more small airports. Because an established VOR has the potential of serving several airports, you'll find that it is the most frequently used approach aid. The surrounding terrain has a lot to say about which airports qualify for approaches using a given VOR, as does distance.

## ✝ Tuning the Receiver

Tuning a VOR is as easy as falling off the bottom wing of a Stearman; it's simply a matter of making the proper numbers appear in the windows. But make it an unfailing habit to *identify* the station after tuning—it can ruin your whole day when you fly a beauty of an approach, only to run into a mountain because you didn't have the right station tuned. The FAA is careful to assign frequencies so that only one VOR can be received during an approach, except when Murphy's Law comes in to play: "If there's a way for something to go wrong, it will." *Tune and identify* is a rule that will keep you out of trouble.

So you're not a Samuel F. B. Morse, and maybe the only code you can really understand is SOS. Fear not, the coded identifiers are transmitted slowly enough for you to follow them, matching what you hear with the dots and dashes on the chart. After a while, close-to-home stations will become familiar patterns of dits and dahs. And if you want to become more professional about instrument flying, there are a number of quick-learning techniques for becoming fluent in Morse code.

## ✝ Initial Approach

By turning to the inbound heading and holding it, you will soon be able to tell whether you are drifting, and which way; the rate at which the CDI moves away from center gives you an idea of how *fast* you are drifting. As soon as the needle moves, start bracketing the course, as outlined in Chap. 9. The secret is to recognize drift, turn to a heading to correct it, and hold the heading until something happens. Now, your technique depends on the direction and velocity of the wind; store the correction in a corner of your memory so you can anticipate headings and groundspeeds during the approach.

## ✝ Station Passage and Tracking Outbound

You've done a good job of tracking to the station when the TO-FROM indicator reverses itself in the blink of an eye; the faster it flips, the closer you are to the station. But even if it takes several seconds, you have officially passed the VOR as soon as FROM appears, and it's time to proceed with the next segment of the approach.

The best way to start is with the four Ts: *Time, Turn, Throttle, Talk.* Start your stopwatch or note the *time, turn* to the outbound heading, *throttle* back to slow down and descend to the procedure turn altitude, and finally, when everything's under control, *talk*; let ATC know that you are outbound from the VOR. Don't be in a rush to make this report; the controller cleared you for the approach and knows what you're doing, so get the airplane going in the right direction before you pick up the mike.

You are required to get on the outbound course before starting the procedure turn, so after station passage, put the correct number on the OBS and take a look at the CDI; it will tell you which way to turn.

You've been descending all this time, of course (gets a little like rubbing your head and patting your stomach, doesn't it?), and your target is the procedure turn altitude, on course, ready to begin the turn. A VORTAC (VOR station with distance-measuring capability) makes it a simple matter to determine your distance from the station. In the absence of DME, keep track of the time outbound.

# ✞ Procedure Turn

Procedure turn altitude is a minimum to be observed until you are on course inbound, at which time you can usually descend. (Sometimes the procedure requires that you maintain that altitude all the way to the FAF.) During the turn, change the OBS to the inbound course. As the CDI begins to center, start your turn to the inbound course, applying drift correction as necessary. (Detailed techniques for the various types of procedure turns are found in the first section of Chap. 15, Instrument Approaches.)

# ✞ Postprocedure Turn and Final Approach

Unless obstructions make this airspace untenable, the postprocedure turn course to the station will be lined up with the final approach, that electronic line down which you're going to fly to the minimum altitude. Take advantage of this opportunity to really nail down the drift correction; you can get a general idea which way the air is moving as you fly this segment. You should also be slowing the airplane to approach speed (an important part of setting up a stabilized approach) and accomplishing the before-landing checklist. Having these things done early will let you focus your attention on precise flying during the final approach segment.

The closer you get to the station, the more the CDI seems to defy your efforts to keep it centered. The needle will move more frequently and rapidly, but think of this as increased accuracy (which it really is), and make the most of it with very small corrections. When the CDI begins to move, start bracketing; a few "cuts and tries" will establish a heading just right for the wind that exists. Here's where a needle chaser will turn the airplane 30 or 40 degrees, whereupon the CDI comes roaring back across the center and off to the right, prompting a 2g turn the other way, which then makes the needle zip to the other side. While this is happening, the airplane is closing on the VOR, which makes the CDI even *more* sensitive. By now, fixation has set in, and the needle chaser has lost all interest in altitude and airspeed. As if that weren't enough, Approach Control will always choose this moment to ask, "Barnburner 1234 Alpha, what are your intentions after this approach?" The reply to that question is frequently unprintable!

The TO-FROM reversal will be abrupt over the VOR, and it's your cue to exercise the four Ts again. You should report over the final approach fix inbound, but all you need to say is, "34 Alpha, VOR inbound."

The final approach segment is the climax of your short love affair with this VOR station; everything depends on your making the right moves. Your attention should be devoted to: (1) keeping airspeed, altitude, and heading right where you want them; (2) descending on course at a rate that will get you to the MDA well before time runs out; (3) keeping track of the time elapsed so you'll know when to go around should it become necessary; and (4) looking for the airport. Don't expect to see it, and you'll get a nice surprise when it comes into view.

Give yourself a break by getting down to the MDA as soon as practical because you'll likely establish ground contact that much sooner. Remember that as groundspeed increases, rate of descent must increase if you are to lose the same amount of altitude in a given distance.

A typical VOR approach (excluding those situations where unusual terrain features and/or distance from VOR to airport require higher minimums) will bring you to the missed approach point 300 or 400 feet above the airport elevation and will require visibility of a mile or more. Keep the CDI centered, descend to MDA, fly out the time, and if the visibility is adequate, you'll find yourself all set up for a normal landing.

# ✛ VOR Approach with DME

Distance-measuring equipment (DME) adds a second dimension to the capabilities of a VOR station. Without DME, the best you can do is locate yourself somewhere on a radial, dependent on timing to provide information relative to distance. Considering changes in wind velocity and airspeed, and the interpolation required on the time-distance tables, your position on a VOR approach is a good estimate at best. With DME, the added precision usually results in more favorable minimums for the approach.

Approaches using DME fall into two operational categories: those in which DME provides step-down fixes and lower minimums than the VOR-only procedure and those in which DME is an absolute requirement. In the first case (DME optional), a pilot experiencing DME loss could continue the approach, but at the higher VOR-only MDA. The second category requires *both* VOR and DME equipment, and it makes no provision for pressing on without DME.

Check the approach chart to determine whether DME is required or is considered an adjunct to the procedure. The key is the inclusion of a separate set of minima for DME-equipped aircraft, indicating one MDA when you've got it, another when you don't. In Fig. 17-1, you are permitted to descend to 1020 feet MSL until the 7.5-mile fix (ALEXE) is identified and then you may go down to the MDA of 600 feet. If DME is not available, you must maintain 1020 feet until reaching the missed approach point; DME has in effect bought you 420 feet. If you use such an airport frequently, DME might pay for itself in one season by permitting landings from approaches that otherwise would have been missed.

On the other hand, a VOR-DME procedure that does not offer the option will have only one set of minimums listed, and that approach *cannot be executed* unless you are equipped with *both* VOR and DME receivers in working order (assuming that both transmitters are also operative).

The chart for Cheboygan, Michigan, illustrates still another means of putting DME to work for a more efficient approach procedure (Fig. 17-2). Instead of maneuvering your way through a procedure turn, you may be cleared to intercept the 8-mile DME arc, proceed around that circle until reaching the final approach course, and then fly

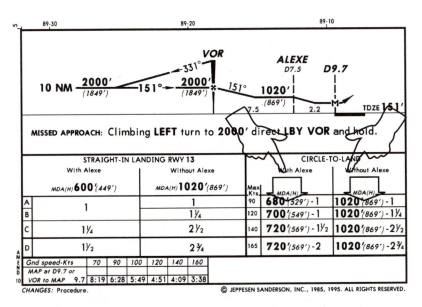

**Fig. 17-1.** *With DME available for more accurate positioning on the final approach course, the MDA is considerably lower. (Copyright Jeppesen Sanderson Inc., 1985, 1995; all rights reserved; not to be used for navigation)*

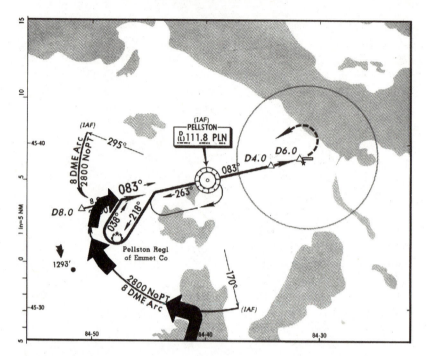

**Fig. 17-2.** *"Intercept and proceed via the 8-mile arc, cleared for the approach." Note the minimum altitude and NoPT on the arc. (Copyright Jeppesen Sanderson Inc., 1985, 1991; all rights reserved; not to be used for navigation)*

straight in to the airport. Although it requires a bit more technique and some diligent practice before it can be accomplished smoothly, flying an arc approach can save you time and give the controller additional flexibility in spacing inbound traffic.

For example, suppose you are inbound from the south and have been cleared to "proceed inbound on the 190 radial until intercepting the 8-mile arc, cleared for the VOR Runway 9 approach." Nothing different about it so far, but when the DME indicator gets close to 8, be ready to make your move. Remember that you are approaching on a 90-degree intercept heading, and things will happen rapidly when the time comes; here's where practice comes booming through in spades, where you should know what mileage lead to use. For groundspeeds of 150 knots or less, a lead of 0.5 miles should be used; in other words, start turning when the DME reads 8.5 miles. As groundspeeds increase, you can compute the proper lead by using this formula: 0.5 percent of your GS = lead in nautical miles. A standard-rate turn is assumed in both situations.

It is obvious that you must turn left 90 degrees, but once you're on the arc, don't attempt to fly in a continuous circle to keep the DME on 8; that would require your complete attention to the detriment of all the other things that are going on. Your flight path should be a series of segments that are tangential to the arc, and it is convenient to use 20-degree heading changes as a guideline. When you complete the turn to 280 degrees, the DME should read very close to 8, and it should stay there for a short while, depending on the wind and your airspeed. (Don't forget to set the OBS on 083 as soon as you complete the turn.) If a west wind drifts you toward the station, the DME may move toward 7 miles, but will eventually slide back to 8 as you fly into the arc again.

In a no-wind situation, you should turn to the right 20 degrees (new heading of 300 degrees) as soon as the indicator starts to move away from 8 miles. Hold this heading until you once again observe the mileage increasing then turn another 20 degrees to the right. Keep it up, turning further and further to the right to keep the DME as close to 8 as possible, until you notice the CDI beginning to center. You are on another 90-degree intercept, so keep turning toward the inbound course to center the CDI. From here on, it's just another VOR approach. Fly straight down the chute to the missed approach point. A procedure turn is neither desirable nor permitted, since you are already lined up on the final approach course.

Pilots whose instrument panel sports a Radio Magnetic Indicator (RMI) with a needle slaved to the VOR instead of, or in addition to, the ADF have one more thing in their favor when cleared for an arc approach; it takes most of the work out of staying "in orbit." At the completion of the first 90-degree turn, the RMI will point to the right wing-tip position (or the left, depending on the situation) and will immediately begin moving toward the tail, unless you're flying head-on into a hurricane. When the RMI pointer has moved tailward 10 degrees, make your first 20-degree heading change, observing the DME to confirm that you're still close to the 8-mile arc. In a light wind, the 20-degree turn will put the RMI pointer 10 degrees ahead of the wing-tip position, and you can wait until it returns to 10 degrees behind before changing heading again.

# 18

# The Instrument Landing System Approach

This is the granddaddy of them all when it comes to getting down close to the ground, and with the most sophisticated equipment, the point can be right on the runway—a zero-zero approach. Special authorization for airport, airplane, and pilot is required for Category II and III (reduced minimums) ILS approaches, but a typical Category I procedure, available to all adequately equipped aircraft and any instrument-rated pilot, allows landings from a Decision Height of 200 feet if visibility is one-half mile or better. *Caution:* Some approaches have DHs higher than 200 feet and visibility minimums greater than a half mile; check each approach chart you use and know what you're getting into.

In strict legalese, an ILS approach may be conducted only if all four electronic components of the system are operating and receivable in the aircraft. These four—localizer, glide slope, outer marker, and middle marker (defined in Chap. 2, The Language of Instrument Flying)—are sometimes supplemented by an Approach Lighting System (ALS). The important thing to understand is that you may not descend to "full ILS" minimums if your aircraft is not equipped with all the black boxes, if all of them aren't working properly, or if the transmitters on the ground are out of business. The localizer is the backbone of the ILS, and when it's inoperative, you might as well request some other procedure; without a localizer to line you up with the runway, the approach is impossible.

Lack of an approach lighting system raises the minimums a bit, losing (or not having) an outer marker receiver bites even deeper, and without a glide slope the approach becomes a nonprecision procedure, which increases the minimum altitude and visibility requirement significantly.

## ✝ Flying the Localizer

It's the same receiver and the same indicator (the CDI) you use for VOR navigation, but some important electronic changes take place when you select a localizer frequency. First, the CDI becomes more sensitive because the localizer course is only about 5 degrees wide, meaning that the needle moves much more rapidly and the displacements are much larger (the CDI is roughly four times as sensitive as it was when tuned to a VOR). Second, the Omni Bearing Selector (OBS) with which you select VOR courses for navigation is cut out of the system; the CDI is now responsive only to a left-of-course or right-of-course signal or a blend of the two, which gives you an on-course indication. (Even though the OBS is inoperative, it's a good idea to set it on the localizer course as a reminder of the track you want to maintain on the approach.)

The instrument designers made it easy for you when they set up the left-right needle; if the CDI *points* to the left, you should *fly* to the left to get back on centerline. When it moves off to the right, correct to the right; it's that simple. The CDI will always show you which way to turn to get back on course when you are flying *toward the runway on the front course* (during a normal approach) or *away from the runway on the back course* (executing a missed approach). You will always fly *toward* the needle on an ILS approach (Fig. 18-1). (Chapter 9 discusses use of the HSI for ILS approaches.)

## ✝ Tuning the ILS Receivers

Because the ILS is a navigational *system,* there is more than one receiver involved. First, set up the localizer frequency on the VOR receiver and listen for the identification; it will always be three letters in Morse code, preceded by the letter "I" (· ·).

When the localizer frequency is selected (always an odd tenth, such as 109.1, .3, .5, .7, .9), it will automatically tune the glide slope

**Fig. 18-1.** *Whenever the heading is the same (or nearly the same) as the front course of the localizer, a correction toward the needle will return the aircraft to the centerline (aircraft #2 and #5). When headed in the opposite direction (aircraft #3 and #6), the pilot must turn away from the CDI to get back on course. Aircraft #1 and #4 show "on course" indications.*

receiver, a completely separate unit. Because there is no way you can identify it aurally, don't worry about it. If you are reasonably well lined up for the approach, the localizer and glide slope needles will come alive when the frequency is selected. Don't panic if the needles refuse to budge when you're approaching from the side; the ILS signals are somewhat directional in nature and are not very strong or reliable except on the approach course, where they are intended to be used.

Tuning the next navaid is not really tuning at all because the 75 megahertz marker beacon receiver is a single-frequency, non-tuneable radio. *Every* outer marker and *every* middle marker transmit a signal on 75 megahertz, the only difference being in the pattern; the OM sends out continuous dashes (— — — —), while the MM transmits alternate dots and dashes (· — · — · —). Make sure that the receiver is turned on and that *low* sensitivity is selected. Check the blue (OM) and amber (MM) marker beacon lights if you have them; they're not required, but will help you identify the markers as you pass overhead.

Although not a requirement or official component of the ILS, you will find compass locators (LOM) installed at many airports. The LOM is an NDB colocated with the OM, providing navigational guidance to the OM (that's why it's called a "locator"). It also serves as an additional indication of *passing* the outer marker, and you may legally substitute ADF for the marker beacon signal if it's not available. (A radar position is also acceptable.) The compass locator is tunable, with its frequency (from 190 to 415 kilohertz) and a Morse code identifier printed on the approach chart.

# ✛ Initial Approach

A terminal with enough traffic to justify an Instrument Landing System will probably be radar equipped, which means that the initial approach will consist of vectors until you are established on the localizer. Because of this, procedure turns on an ILS approach are infrequently required, but there may come a time when you have to accomplish the approach all by yourself with no help from radar, so you should know how to do it.

Notice that the ILS Runway 27 Approach Chart for Benton Harbor, Michigan (Fig. 18-2), indicates two points from which flights are normally cleared to the LOM to begin the ILS approach. Each of these initial approach routes has its own course to the LOM, the distance, and a minimum altitude; from the MAPER intersection, it's 076 degrees, 19.8 NM, and the lowest safe altitude is 2400 feet; from the GIJ VOR, the course is 001 degrees, 21.5 NM, and no less than 2400 feet.

Some initial approach routes are nearly lined up with the localizer course and do not require a procedure turn. When this situation exists, the route will be labeled "NoPT," and you are expected to proceed as charted, straight in, with no procedure turn. An example is the route shown from the Keeler VOR: The minimum altitude (2300) coincides with procedure turn altitude, and you should be ready to

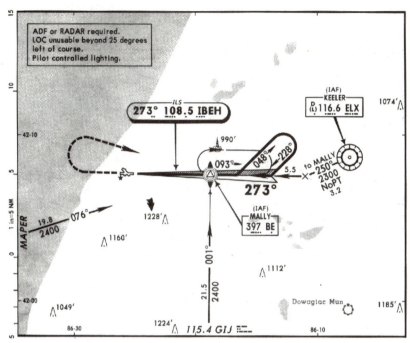

ADF or RADAR required.
LOC unusable beyond 25 degrees
left of course.
Pilot controlled lighting.

273° 108.5 IBEH

(IAF)
KEELER
ⒹⒾ116.6 ELX    1074'

990'
093°    048°    228°
273°    to MALLY
X—250°
2300
NoPT
3.2
5.5

(IAF)
MALLY
397 BE

1228'
1160'
076°
MAPER
19.8
2400
1049'
1224'
1112'

Dowagiac Mun    1185'

21.5
2400

001°

115.4 GIJ

**Fig. 18-2.** *Approach procedure chart for the ILS Runway 27 procedure at Benton Harbor, Michigan. (Copyright Jeppesen Sanderson, Inc., 1986, 1995; all rights reserved; not to be used for navigation)*

commence descent upon intercepting the glide slope. The controller should have you down to a compatible altitude by the time you cross Keeler, but if you need a procedure turn to lose altitude (ear problems, maybe), by all means ask for it, and your request will most likely be granted. But don't do it without a clearance because the controller may have a faster aircraft right behind you. Another situation that precludes a procedure turn is one in which you have been radar vectored to the final approach course (the localizer) and have been cleared for the approach; in this case, you're all lined up, courtesy of radar, and there's no need to turn around.

Assume you have been cleared "from over the MAPER intersection direct to MALLY, maintain 4000 feet, cleared for the ILS Runway 27 approach." You're on your own now, and after acknowledging the clearance, begin the descent to 4000 feet and track inbound to MALLY via the 076-degree course. You should be tuning radios, cleaning up the cockpit, listening to other aircraft, and generally getting your ducks in a row for the approach. If you have two VOR receivers, tune #1 to the localizer frequency and check the identification; tune #2 to the GIJ VOR and set the OBS on 001 as a cross-check. Use the same receiver for approach work every time, and you'll spare yourself the embarrassment of flying the wrong needle, to say nothing of the hazard involved.

The CDI will be a long time moving because the signal is so narrow, but as you get closer to the outer marker, the CDI will come alive, your cue to start turning to the outbound heading of 093 degrees. With practice, you'll develop the proper lead to prevent overshooting the localizer course; it's a function of airspeed, rate of turn, and wind component, all three of which must be taken into consideration.

A Beechcraft Baron pilot (and his passengers) found out the hard way that an instrument pilot is expected to find his own way around the approach procedure in the absence of radar guidance. Although this accident occurred hundreds of miles from Benton Harbor, Michigan, the layout of the ILS procedure is almost identical and will help you visualize the situation.

The Baron was approaching from the northwest (roughly a 45-degree angle to the localizer) and had been requested by the Center controller to descend rapidly from cruise altitude in order to expedite traffic flow into the airport. The pilot replied, "Okay, we'll drop it out," and he did just that. Radar plots showed 1500 feet per minute at more than 200 knots. Just a mile or so from the marker, the Center controller cleared the Baron for the ILS approach and asked the pilot to contact the tower.

With the weather hovering at 200 feet and 1 mile visibility, it would seem important to take every opportunity to make the first approach a good one, but upon crossing the marker, this pilot—for reasons unknown—turned *to the right,* toward the airport. The tower controller (without radar and expecting that the pilot would proceed eastbound for a procedure turn) asked for a report at the marker inbound, and a couple of minutes later, the pilot replied, "I'm at the marker, inbound."

Sounds all right; that's a little fast, but not an unreasonable amount of time for a sharp pilot in a hurry to get turned around and headed for the airport. Unfortunately, because of the high speed and the right turn, the airplane was very close to the *middle* marker and well on its way down to Decision Height. The pilot wandered back and forth across the localizer, either completely disoriented or maybe looking in vain for the runway, which was now well out of sight. The low-altitude flight continued to a point 7 miles west of the airport, where the operation was brought to an abrupt halt by the guy wires of a television tower.

Lessons to be learned:

- If you can't handle an expedited approach, *turn it down.*
- When weather conditions are close to minimums, *fly the complete approach, as published.* Give yourself a chance to get it right the first time.
- Figure out well ahead of time which way you'll need to turn at the marker and follow through with that plan when you get there.
- Whenever you are not *absolutely certain* of the meaning of a clearance, request a clarification.
- When things "just don't feel right" on an approach, they probably aren't. Execute the missed approach procedure and come back for another try.

# ✝ Station Passage and Tracking Outbound

A pilot flying a fully equipped airplane has no excuse for missing station passage at the outer marker because of all the indications. The most obvious will be the strident blare of the marker's continuous dashes, while at the same time the blue light is flashing and the ADF needle swings from nose to tail. Once again, it's time for the four Ts.

Most controllers will clear you to procedure turn attitude well ahead of time, but if that doesn't happen, plan to lose at least half the altitude outbound, the rest inbound, and make your descent at a practical yet comfortable rate. Keep the welfare of your passengers in mind; for most nonpilots, descending faster than 800 to 1000 feet per minute can cause problems.

# ✝ Procedure Turn

Still referring to the ILS procedure in Fig. 18-2, turn to a heading of 048 after 2 minutes outbound, and the procedure turn is under way. When it's time (depending on the type of procedure turn you elect), turn right to 228 degrees, which sets you up for intercepting the localizer once again. As your heading approaches 228, watch the CDI out of the corner of your eye; if it's beginning to center, continue your turn and you'll roll out on course. If at 228 it hasn't moved from the left side of the instrument, roll out on 228 and hold it until the CDI starts moving; the proper lead comes with practice. In the opposite situation, the CDI starts moving toward center during your turn inbound; obviously the result of a northerly wind component, it will require an increase in the rate of turn so that you won't overshoot the localizer.

The time from localizer interception to the outer marker is a golden opportunity to find out what heading will keep you on course; you can get a good idea of what's going on windwise right here.

# ✝ Postprocedure Turn and Final Approach

Inbound to the airport, the CDI is making sense again—turn left when it's off center to the left, and vice versa—and it's time to configure the airplane for the final approach. As your experience, ability, and confidence increase, it's good practice to keep your airspeed somewhat higher than the final approach speed until you need to slow down to get stabilized. (At busy terminals, you will often be asked to maintain *cruise* airspeed to the outer marker, but don't do it if the resultant "too many things at once" is more than you can handle; tell Approach Control you can't comply, and let *them* handle the traffic separation problem.) By setting up the airplane in approach configuration at approach airspeed, you will need only to reduce power to establish a rate of descent that will keep the glide slope needle right where it belongs.

Still inbound to the OM, maintaining 2300 feet (see Fig. 18-3), note the glide slope indicator begin to move from its full UP position; this is a good indication that you are very close to the marker. If you are carrying some extra airspeed, it's time to bleed it off; the final scene in this drama is about to commence. As soon as the glide slope needle centers, reduce power (or lower the wheels) and begin the descent, even though you have not yet crossed the marker. It's completely legal and is an important part of the approach procedure. Check the profile view of the approach and you'll see that, if the glide slope is followed, you will cross the OM at an indicated altitude of 2203 feet. A glance at the altimeter will provide an accuracy check before continuing the approach; if it's way off (how much is up to you), you should have second thoughts about continuing the approach.

On course and on glide slope, the marker indications will be just as numerous as they were outbound, and at station passage, only two Ts apply—*Time* and *Talk*—"Barnburner 1234 Alpha, marker inbound." That's all you need to say, for two reasons: First, it's merely a confirmation because the approach controller has already informed the tower controller that you're on the way; second, you are busy flying the airplane, and you really haven't time to waste carrying on a conversation. Likewise, you shouldn't be concerned if you don't (or can't) make this report right away; your primary job is controlling the airplane, tracking the localizer and glide slope as accurately as you can. Tower will usually come right back with clearance to land; they've been expecting you.

And down you go; down, down, down—from outer marker to runway-in-sight or missed approach can be a long couple of minutes. Make slight power corrections to keep the glide slope needle centered—it's always directional; down means fly down, up means fly up—and make small heading changes to keep the CDI centered. Don't try to "fly" the localizer and glide slope; you'll inevitably chase them back and forth and up and down, and you can't win because the needles can move faster than the airplane.

The only technique that works well is to include the CDI and glide slope as additional instruments in your cross-check. Make corrections on the *attitude* instruments to put the ILS needles where you want them to be. If the CDI slides off a bit to the right, turn to the right 5 degrees (just press on the rudder; don't try to bank the airplane for a 5-degree turn) and watch to see what happens; it's important to pick a corrective heading and *hold it* until you can determine the effect on the CDI. Still sliding right? Turn another 5 degrees and see what that does. When you finally get the needle centered, take out the corrections in small increments with rudder only, and before long you'll have a heading that will maintain the localizer course. The same technique is used for vertical corrections; increase or reduce power *slightly* to change the rate of descent so that you will fly back onto the glide slope.

Down some more until, with localizer and glide slope needles centered, you see 887 feet on the altimeter, assuming you have obtained a local altimeter setting. You are at the Decision Height (DH) and you've got to make up your mind—either fish or cut bait! If the runway environment (approach lights, runway lights, runway markings) is visible, and if you are in a position to make a normal landing (that's up to you), you may proceed to touchdown; that's the "fish" situation. But if those conditions are not met, you'll have to "cut bait" and execute the missed approach procedure. On this point, the regulations are brief, concise, and clear.

There will be very few times when you start an ILS approach and don't complete it. For one thing, the weather doesn't get down to ILS minimums and stay there that often; furthermore, you will probably not be out flying when the weather is that bad. Especially in a single-engine airplane, you are pulling the string all the way out in such conditions. If you should experience that increasingly rare occurrence of power failure, you've nowhere to go but down, and because the glide slope is rather flat (typically a 3-degree slope), you'll need power to get to the runway. Without power, you'll likely land at some point other than on the runway.

If you spot the runway from a half mile out, which is the typical visibility minimum for an ILS approach, you should be able to make it the rest of the way. (When you're planning an IFR flight to an ILS-equipped airport, don't count on one-half mile. The visibility minimum for Benton Harbor, Michigan, as shown in Fig. 18-3, is three-quarters of a mile. Always check the chart for your destination.) The law is equally clear regarding your actions if you lose sight of the airport after you go below Decision Height. You must execute an immediate missed approach—no hesitation, no hunting around for the field; get out of there, *now!* Come back for another try if you like, but don't press your luck if the runway doesn't show up at Decision Height. (When you are cleared for the ILS approach, then to circle for landing on some other runway, you must observe *circling* minimums, which will limit you to a minimum descent altitude [MDA]. You *may not* go down to the DH, even though the airplane is fully equipped and all the ground-based transmitters are working properly. The published Decision Height is intended for only *one* runway.)

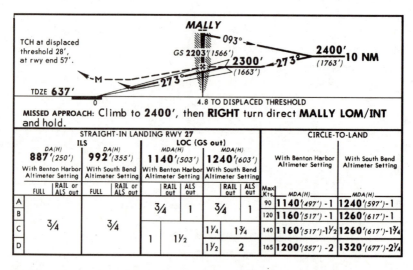

**Fig. 18-3.** *Profile view and minimums for the ILS Runway 27 procedure at Benton Harbor, Michigan. (Copyright Jeppesen Sanderson Inc., 1986, 1995; all rights reserved; not to be used for navigation)*

Measured above the highest elevation in the touchdown zone (usually the first 3000 feet of runway), Decision Height is accompanied on the approach charts by another figure called Height Above Touchdown (HAT). A typical ILS procedure puts DH 200 feet above the runway, but sometimes it will be considerably higher because of terrain or obstructions. For example at Roanoke, Virginia, the DH is 500 feet above the runway, and the visibility minimum is 1.25 miles because low clouds around Roanoke are *cumulus granitus* (that is, full of rocks)!

# ✝ The Localizer Approach

Some general aviation aircraft are without glide slope receivers, and some airports have only a localizer transmitter installed. In addition, glide slope transmitters and receivers have been known to develop malfunctions. Any of these circumstances dictate a localizer approach, which means that everything is there to help you except the vertical guidance of the glide slope. This automatically becomes a *nonprecision approach,* with attendant higher minimums. You will be limited to a Minimum Descent Altitude (MDA), and you must rely on timing (or DME on occasion) for determination of the missed approach point.

A localizer approach is therefore flown very much like a VOR approach (it's up to you to begin descent at the appropriate time, get to the MDA, and determine the missed approach point all by yourself), with the increased accuracy of the localizer to line you up with the runway more precisely. The Final Approach Fix on a localizer approach is frequently a marker or an NDB, but it can also be a VOR intersection, bearing to an off-course radio beacon, or a DME fix when a VORTAC is located on or very near the airport or when the localizer incorporates DME. On occasion, the approach chart carries the notation that two VOR receivers are required: one for the localizer and the other to determine the intersection making up the Final Approach Fix. If the FAF depends on an NDB or a bearing to a nearby radio beacon, you'll also need an ADF receiver.

Initial approach, procedure turn, and final approach are flown just like the full ILS, with one exception: Plan to arrive at the MDA well ahead of the missed approach point to provide more time to look for the runway.

Controllers won't ask if your airplane is properly equipped; they'll simply clear you "for the runway so-and-so ILS approach." If you don't have a glide slope receiver, it's up to you to realize that you cannot execute a precision approach, and you must look up the "Localizer Only" numbers on the chart. (You are also required to let the controller know that this approach will be "Localizer Only.") Don't ever attempt to fly down to a Decision Height unless you are receiving a usable glide slope signal; that's really asking for trouble. And don't always assume that a localizer approach is authorized for each ILS procedure. On some approaches, the angle required to clear terrain is so critical that the full procedure is required—no glide slope, no approach.

"No glide slope" applies to equipment on board the airplane as well, illustrated by the unfortunate outcome of this approach: Having purchased a used airplane on the west coast, a noninstrument pilot found himself stranded halfway home by inclement weather. He was anxious to be back for Christmas and hired an instrument instructor to fly with him, expecting to gain some IFR instruction along the way.

The CFII was very recently qualified, the ink barely dry on his certificate, and he had zero experience in this airplane type, a Cherokee Six. Having satisfied himself that all was well with the airplane and its equipment (some avionics work had been performed as a condition of the sale in California), the instructor and his passengers departed, and the flight was unremarkable until late in the ILS approach at destination. It was long since dark, the weather was near minimums, and the last thing the CFII remembered (he was the only survivor) was a glimpse of the approach lights just before impact. The airplane crashed 1 mile short of the runway.

In a postaccident statement, the instructor described his final approach, with localizer and glide slope needles centered all the way, airspeed at a constant 90 knots, rate of descent 500 feet per minute. What he *didn't* take into account was a notation on the avionics work order; "G/S inop"—glide slope inoperative.

In nearly all light-airplane installations, an inoperative (or failed) glide slope indicator will go to a centered position and call up a red warning flag. That's good, but an unwary pilot—especially one sitting in the right seat—could ignore the flag and conclude that the airplane was on the glide slope. Although a 500-foot-per-minute descent is just about right for a 3-degree glide slope at 90 knots, keep in mind that the proper rate of descent is dependent on *groundspeed,* not airspeed. The faster you move across the ground, the faster you must move vertically in order to stay on the glide slope.

Obviously, it would have been impossible for this airplane to hit the ground *if the pilot had stopped the descent at Decision Height.* A steadfast refusal to go below DH *regardless of other instrument indi-cations* unless and until the runway is clearly in sight can prevent problems such as this one.

## ✝ The Localizer Back-Course Approach (LOC-BC)

One statement—"a back-course approach is the same as a front-course approach without the glide slope"—should suffice to explain this pro-cedure, but because the CDI appears to work in reverse, many pilots develop hang-ups about back-course approaches.

A back-course procedure always lines you up with a specific run-way and offers lower minimums than either VOR or ADF. It's up to you to determine by timing when you have reached the missed approach point, and the final approach fix is frequently an intersection instead of a marker or compass locator. Other than that, the back-course approach is as easy and as accurate to fly as the front course.

Initial approach and procedure turn techniques are the same as before, so this discussion begins after you have turned around and are flying toward the runway on the back course, and just before reaching the final approach fix. Since completing the procedure turn or intercepting the localizer, you have been flying away from the nee-dle to effect your course corrections. The technique of making a heading change and *holding* it until you see a change on the CDI is just as relevant as it was on the front course, and as you approach the FAF, get the airplane ready to begin descent. When the marker or intersection is passed, three Ts apply: *Time, Throttle,* and *Talk*—you owe ATC (usually the tower) a report.

The LOC-BC approach is subject to the same regulations as all the others regarding MDA, landing conditions, and missed approach pro-cedures. If you can get to the runway using visual cues and normal maneuvering, go ahead and land. If these conditions don't exist or if they can't be maintained after leaving MDA, you've no choice; execute the missed approach.

Flying a back course approach properly is more a state of mind than anything else. The procedures, principles, and restrictions are not that much different than any other approach; you've just got to sit there and ignore the snickering from the back seats as you mumble to yourself "fly away from the needle, fly away from the needle, fly away from the needle."

# 19

# High-Altitude IFR

The turbocharger, that marvelous pumper of air that makes reciprocating engines think they're running at sea level all the time, has literally boosted lightplane operation into volumes of airspace previously denied to all but the jets and turboprops. Until small, relatively inexpensive turbochargers became available, the practical altitude limit for lightplanes was 10,000 feet, and most of them spent so much time getting up there it was hardly worth the effort except on long trips or when flying over high terrain. The single-engine ceiling of normally aspirated light twins left more than a little to be desired; in essence, when you finally got to the higher levels, there just wasn't much power left.

But now you can "slip the surly bonds of earth" behind a supercharged engine or two, and climb into the high-altitude world of crystal-clear air. Up where altitudes change to flight levels, where the winds often move faster than a J-3 can fly, you look down from your oxygenated perch at all the murk below and wonder why you didn't do this sooner. And there's almost as much power available up here as you had at takeoff, which makes for much-improved single-engine performance in light twins. (If the engine quits on a supercharged single, you're still faced with a forced-landing problem, but you'll have a lot more time to solve it.)

A wealth of benefits awaits the pilot who shells out the extra dollars for supercharged engines:

- Flight above most of the weather, especially that troublesome twosome, turbulence and icing.
- Much faster climb through the weather to "on top."

- The ability to see most thunderstorms and avoid them.
- Higher true airspeed at altitude.
- Much greater latitude in selecting the best altitude for winds or weather.
- You can take advantage of fantastic tailwinds.
- When it's clear up there, it's *really* clear; no problem seeing other aircraft.
- Comparatively little traffic, and everyone else in this airspace (above 18,000 feet) is under positive control.
- The ride will almost always be smoother than at lower levels.
- If you have obnoxious passengers on board, you can slyly unplug their oxygen hoses and watch them go to sleep.

And the list goes on, with new turbopilots adding their own pet joys. But like anything worthwhile, all this has a price (in addition to the cost of the equipment) that must be extracted in the form of additional pilot knowledge, new techniques to be learned, and increased vigilance. You'll have to become familiar with a whole new set of enroute charts if you're going into the high-altitude route structure because the low-altitude paperwork won't do the job above 18,000 feet; it's like trying to take a trip on interstate highways with a county road map. Some new techniques will crop up, like learning how far out to request descent or getting accustomed to high-speed descents. When you haul people to altitudes that might affect them adversely, they have a right to expect you to know all about this new environment and to be the guardian of eardrums, sinuses, and breathing apparatus. High altitudes are not necessarily dangerous, but the hazards are insidious and unforgiving.

# ✛ Baby, It's Cold Outside!

There are several things that you should check more thoroughly than usual as you preflight for a high-altitude trip, and one of these is the cabin heater. On a sweltering summer day, you may look forward to the cooler air of the 5000- to 10,000-foot levels you used before you got the turbos, but when you keep right on climbing into the teens and twenties, you won't want the fresh air vents open! Temperatures remain low throughout the year up there, with the OAT gauge seldom climbing above freezing, even in midsummer. In the wintertime, temperatures of 20 below zero are not uncommon. If the cabin heater fails—or won't start—there's no doubt that you can survive, but preoccupation with keeping yourself warm can trigger a chain reaction of inattention and forgetfulness. Besides, it's just plain uncomfortable to sit in a cold airplane; a simple preflight check of the heater saves you all the trouble. If the heater doesn't fire up to full output when you check it on the ground, take the time to get it fixed or plan the flight for a lower altitude, especially if you have passengers on board for their first flight. No one will walk away from an airplane with as much distaste for the whole business as the one who walks away on numb feet, trying to restore circulation to frostbitten fingers, and mumbling antiaviation epithets through blue lips.

There's nothing like a cup of steaming hot coffee to complete the settling-down process at altitude, after the autopilot has taken over the chores and you can relax for a few minutes. But if you wait until this point to open the jug, be prepared for a real surprise; nearly all thermos bottles are purposely designed with a very effective seal around the cap, and as a result, you will suddenly expose boiling coffee at sea level pressure to the sharply reduced pressure of high altitude. It's a lot like pulling the plug on Old Faithful, with the painful difference of having the geyser right in your lap! No one can do a good job of flying an airplane with superhot boiling coffee all over, and there's always the chance that you may get it in your eyes—'nuff said. Remember to loosen the cap a bit on the way up so that the pressure can equalize. This is a good practice any time, even when climbing to only 5000 feet.

# ✝ What You Don't Know *Can* Hurt You!

You should have a carbon monoxide (CO) indicator pasted firmly and conspicuously on the instrument panel. Some pilots, especially former coal miners, insist on carrying a live canary, but it's much easier to glance at the indicator, which turns dark in the presence of CO. Until now, this may have been of concern only in the wintertime, when heater malfunctions or leaks could fill the cabin with undetectable and very poisonous CO fumes, but you'll be using the heater on nearly all high-altitude flights. Have something on board to let you know when a problem exists.

# ☩ Oil Is Important

The engine oil supply is another preflight area not adequately served by the "kick tire, light fire" concept of preflight inspections. The engines are not only going to be putting out more than normal power in the turboboosted climb to higher altitudes, they will be doing it for a longer period of time and at a relatively low airspeed. Because of this, airplane engines need all the oil specified in the aircraft manual to help dissipate the thermal energy that is the inevitable result of higher power settings. When you push the nose over at altitude, the increase in airspeed and the well-below-zero air whistling by the cylinders will improve the cooling situation, but during the climb, the oil coolers help get rid of a lot of BTUs. Give your engines a break by making sure they have enough oil before you start a flight. On some installations, the oil supply for the turbo is a separate system and must also be checked; it too is responsible for heat transfer in addition to its lubrication chores.

# ☩ Take Along Something to Breathe

Check the oxygen gauge for full pressure as you continue the specialized walkaround for a high-altitude flight. You should keep this system filled for a number of reasons, not the least of which is always having oxygen when you need it. How frustrating to climb above an area of icing or turbulence, only to be forced back down where you don't want to be because you didn't fill the oxygen bottle. And when you *do* refill it, be sure that nothing except *aviator's breathing oxygen* goes into the system. Anything else—even medical oxygen—may contain water vapor, which has a nasty habit of freezing in the lines and shutting off the flow completely. You may save a couple of dollars buying lower grade oxygen, but you'll lose the whole shooting match when the system freezes shut.

One parting shot about preflighting the oxygen system; plug in the passengers' masks and put them where they can be seen so that no one gets a big surprise when you break out the nose hoses on the way up. Some folks just don't take kindly to sticking their faces into a big plastic nose with a balloon on the end, and it's much better to find out who these people are when you're on the ground and able to reason with them. (Trying to settle down a near-panicked passenger in the back seat when your hands are full flying the airplane can be a horrendous experience.) With masks plugged in, open the valve and assure yourself that all the indicators show oxygen flow; you'll want all passengers to check their supply later on at altitude, but now is the time to make absolutely certain everything is right.

## ✝ Don't Forget the Captain

The extra time spent preflighting the airplane is a complete waste if you neglect the most important system of all—*you*. The nagging cold, the slight headache, can assume monumental proportions when carried to high altitudes; not only are your aches, pains, and bad mood magnified by the environment (in an unpressurized airplane at FL 180, your body is subject to half the atmospheric pressure under which it normally functions), but you will find your basic piloting and navigational skills taxed by the increased speed with which things happen. Here's the situation that may start a snowball effect; less than ideal physical and/or psychological condition coupled with high-altitude effects lead to mistakes and uncertainties, which lead to apprehension and tension, which lead to increased oxygen consumption, and on and on until the worst may happen. If you don't feel up to it, don't fly.

Given the insidious nature of hypoxia, and the fact that each individual experiences slightly different symptoms, the FAA has made arrangements with a number of military aviation facilities throughout the country to provide altitude training for civilian pilots. In a 1-day course, you'll learn more about your high-altitude self than you could learn in years of flying. Every one of the physiological problems is fully explained, and most of them are demonstrated in a carefully controlled environment. In a sense, you can "practice" getting hypoxia and learn to recognize your personal symptoms so that you'll know what to do should they ever show up in the real world. The altitude chamber courses are well done, extremely interesting, and unlike everything else in this business, inexpensive. Write to the FAA's Aeromedical people in Oklahoma City for a schedule, or get an application from the FAA office nearest you.

## ✝ Make Those Turbos Work for You

Planning for a high-altitude IFR flight involves more than just picking out a convenient route, eyeballing the distance, and guesstimating how long it will take. On relatively short flights, you'll discover an optimum altitude for almost any day (excepting really strong winds, icing, or turbulence), and it will probably be higher than you've been flying with normally aspirated engines. With turbos providing the capability for rapid climbs, you'll make money in direct proportion to your altitude, and passengers will be more comfortable in the bargain. It's amazing how people will remember the 10 or 15 minutes in smooth, clear air even though you climbed through a lumpy sky for 10 minutes to get them there. Common sense dictates that you shouldn't go to 15,000 feet for a 15-minute trip, but when you fly high, you'll be more comfortable, safer (because you will be up there where you can *see*), and remarkably close to the elapsed time of the same trip flown at lower altitudes, down in the murk where everybody else is groping along.

On the longer, more critical trips, you want those expensive air pumps to work most effectively for you, so set your sights high. Considerable advantage can be gained even without penetrating Class A airspace (18,000 feet and above), especially when you are headed west. The upper air over the United States usually moves from west to east and sometimes at three-digit speeds, so a general rule would be to fly lower westbound and as high as practical eastbound. It's picking out that westbound altitude that becomes tricky because, even though you may be flying upstream, the increase in true airspeed at altitude could more than make up for the headwind.

You can't go too far wrong using the 2-percent-per-1000 feet rule of thumb to help make the altitude decision. If your airplane indicates 150 knots at 10,000 feet, you can figure on a true airspeed of close to 180 knots; this will hold up as long as the turbos can maintain constant power. The difference becomes quite dramatic at high altitude, where true airspeed may often be half again what the indicator shows.

Of course, the price to be paid is the time you spend climbing to altitude at a high power setting and a relatively low airspeed. How far you intend to go becomes important, too; the extra knots generated by the turbos may not be justified by the many gallons-per-hour pouring through the fuel injectors at climb power. Sometimes it becomes a matter of deciding whether to go fast for a short distance or to give up some speed in exchange for making the trip nonstop.

Comfort and weather conditions bear heavily on your decision, so there's really only one way to arrive at the best altitude for an extended trip: Get out your computer and weigh performance figures for several altitudes against all the other factors.

# ✈ High-Altitude Charts

It's not as thrilling as breaking through the sound barrier, but when you fly in Class A airspace, you're flying in a new world of regulations. Some major restrictions affect your operations up here, and they should be recognized in the planning phase. For example, you must use the separately published "High-Altitude Enroute Charts" (either U.S. or Jeppesen).

A number of differences are apparent when comparing the high-altitude charts with their below-18,000-foot counterparts; the most striking is that a whole bunch of charts is missing, but you get your money's worth, even though there are only two pieces of pa per in the mail at revision time. Because of the tremendous increase in true airspeed at high altitude, fewer checkpoints are needed, and consequently, the coverage of the charts is greatly increased. Just think how far you can fly without having to break out a new chart! You may also notice right away that the jet routes (even the name "airway" changes up here) have legs much longer than low-altitude airways.

Selection of a route in the high-altitude structure is easy: Pick out the jet route that runs closest to a straight line between here and there and depend on radar to get you started and stopped. Most high-altitude flights get under way with vectors to a nearby VOR or to intercept the desired route. The same situation prevails at the other end, where you will be worked into the approach environment with the all-seeing eyes of radar. Be especially cognizant of Standard Instrument Departures (SIDs), preferred routes (which are published separately for high-altitude operations), and Standard Terminal Arrival Routes (STARs) when filing; you're working with the big-leaguers and should expect to receive the same kind of ATC handling, so be prepared. When you get right down to it (beg your pardon, right *up* to it), route planning for a high-altitude flight is much easier, with fewer checkpoints and more direct legs.

# ✈ Positive Control Airspace

All the airspace at and above 18,000 feet MSL is Class A airspace. ATC exercises positive control in this environment to keep high-speed airplanes from running into each other, and that's got to be good for everybody. To make this happen, some very specific restrictions and requirements have been legislated for all who seek the advantages of flight in Class A airspace.

In the first place, there is *absolutely no VFR allowed*; every flight within Class A airspace must have an IFR clearance with an assigned altitude. The IFR-only requirement also means that you must be instrument rated and current, and your airplane must be properly equipped and inspected. And not just normal IFR equipment either; there must be an operable Mode C transponder installed, above FL 240 you need DME, and you must be able to communicate with ATC on frequencies they assign. For all practical purposes, this means 360 channel capability because you will frequently be requested to "contact the Center now on one two four point five seven" or some other 50 kilohertz-spaced number. (Stand by for 25-kilohertz spacing—*three* numbers after the decimal—it's already in use by some Centers.)

If you insist on flying VFR at high altitude, you can do it legally, but you'll have to go out over the ocean and climb to FL 600, which is the upper limit of Class A airspace.

# ✝ Supplemental Oxygen

An average person will begin to feel the effects of altitude at about 10,000 feet, which no doubt accounts for the longstanding recommendation for oxygen above that level. You as the pilot may not operate an unpressurized airplane between 12,500 and 14,000 feet for more than 30 minutes without supplemental oxygen. Above 14,000 feet, you must use oxygen *all* the time. This rule was not devised to *encourage* nonoxygen flights at those levels, but to *permit* them for short periods to get you over weather and high terrain. (You must provide supplemental oxygen for all occupants when flying above 15,000 feet.)

With turbos, you are on your way to altitudes at which some help in supplying the breath of life is necessary to function properly. And since you probably don't have an "average" person on board, why not plug into the oxygen system at, or soon after, 10,000? It gives you plenty of time to recheck the system before any physiological damage can be done and helps acclimate you and your passengers to the environment in which you'll be operating for the next hour or so.

Make sure that everyone on board understands the importance of remaining on oxygen until you give the word to unmask on descent. Now open the valve and have everybody check and let you know that they are "on" as evidenced by the individual flow indicators. Most people don't realize how quickly they can join the "blue fingernail set" without oxygen at high altitudes, so make it a practice to check everybody at least every 15 minutes; they'll appreciate your concern. The effects vary with individuals, but turbo altitudes without oxygen can produce the symptoms shown in Table 19-1.

Small children represent a special liability at high altitudes in an unpressurized airplane, and very wee ones shouldn't be exposed at all to oxygen-requiring flight levels. Allowing passengers to sleep while at altitude can hardly be avoided, but make frequent checks of their condition.

"No smoking" is absolutely mandatory while oxygen is in use.

Be aware that oxygen and oil can make a very combustible, self-igniting mixture under certain circumstances. Even the oils in some cosmetics—lipstick and oil-based face creams—should be removed before using oxygen. There have been instances of nasty facial burns when stick-type preparations intended to relieve chapped lips have burst into flames in the presence of pure oxygen. Why take a chance?

At the first suspicion of *any* kind of oxygen trouble, descend *now* and do it *fast* so you can troubleshoot at an altitude where the dangers are minimized. The pilot's well-being is more important than anyone else's because if *you* get into difficulty, don't look for much help from a planeload of hypoxic passengers.

**Table 19-1.** Symptoms of Hypoxia

| Feet | Hours | Effects |
|---|---|---|
| 8,000–10,000 | 4 or more | Fatigue, sluggishness |
| 10,000–15,000 | 2 or less | Fatigue, drowsiness, headache, poor judgment |
| 15,000–18,000 | ½ or less | False sense of well being, overconfidence, faulty reasoning, narrowing of field of attention, unsteady muscle control, blurring of vision, poor memory; *you may pass out* |
| Over 18,000 | | Faster onset of above symptoms; loss of muscle control; loss of judgment; loss of memory; loss of ability to think things out; no sense of time; repeated purposeless movements, fits of laughing, crying, or other emotional outbursts |

# ✝ Flight Levels

Forget about altimeter settings as such because at and above 18,000 feet everyone flies at an assigned *pressure attitude*. This means you will set 29.92 in the adjustment window of your altimeter and leave it there until you're out of Class A airspace. You are *flying* at a certain number of feet above the 29.92 pressure *level* (which on a standard day will be at sea level), and so the vice-president in charge of names put the two together and came up with "Flight Level," the designation for assigned altitudes in Class A airspace—shortened to FL 190, etc.

There are going to be times when the 29.92 level is so low (as around the center of a deep depression or "low" in the atmosphere) that 18,000 or 19,000 or maybe even 20,000 feet above it would sag down into the Low-Altitude Route structure and would be unusable. See Table 19-2 for the lowest usable flight levels. When this happens, you will not be assigned a flight level unless it is safely above the low-altitude aviators. Part of your clearance when descending from high altitude will be the altimeter setting in the terminal area; when you descend through 18,000 feet (or the lowest usable flight level), adjust the altimeter accordingly and rejoin the low-altitude crowd.

# ✝ High-Altitude Weather

When you're flying high, weather avoidance doesn't amount to much, except when going through fronts or the hearts of low-pressure systems. There is the occasional problem of getting through some nasty conditions on the way up, but once you're there, most of the weather is below you. Having the capability to operate at the higher levels goes a long way toward all-weather flying, but some days there will be heavy icing and severe turbulence at all levels with solid lines of

**Table 19-2.** Lowest Usable Flight Levels

| Altimeter setting | Flight level |
|---|---|
| 29.92 or higher | 180 |
| 29.91–29.42 | 185 |
| 29.41–28.92 | 190 |
| 28.91–28.42 | 195 |
| 28.41–27.92 | 200 |

thunderstorms lying across every route between here and there; these are the days when it's better "to have flown not at all" than "to have flown and lost."

Icing is not often a problem at high altitudes except over mountainous terrain because clouds that form on high are usually the cirrus type, composed of ice crystals that for the most part bounce off the airplane. But the turbocharged engine does have one drawback in this situation: The ice crystals sometimes collect in the air intakes, partially closing off the supply, and you may experience a significant drop in manifold pressure. You'll be able to limp along at a reduced speed, but your thinking must change as you plan ahead; start checking the weather at terminals this side of your original destination. An alternate air source that can be selected if the intakes ice up completely will help, but you'll pay a price in reduced power. The only way to get rid of the ice is a descent to warmer altitudes.

Turbulence associated with thunderstorms is one thing; clear air turbulence (CAT) is something else. The visible presence of a cumulonimbus operating at full throttle should serve to warn you away; but CAT can claw at you invisibly, when you least expect it, and it may be the surprise of your life. Thunderstorms are rather well forecast, and there's no excuse for blundering into one. Clear air turbulence is not so easily pinned down, although the weather folks are getting better at it every day; when CAT is suspected along your route at the altitudes you wish to fly, take a long second look. When you spot thunderstorms ahead (and from your high-altitude vantage point you can see most of them) ask for a route deviation around them; controllers will seldom turn you down. When it appears that there's no way around and the tops are out of sight, it's time to do something else: Implement plan B. *Always* have a plan B.

The sustained high power and rapid climb capability provided by your turbos will stand you in good stead in an icing situation *if you act promptly*. Two sets of circumstances usually prevail; you may be faced with getting through an icing layer while climbing to clear air or too-cold-to-freeze temperatures, or you may find yourself picking up ice unexpectedly in a cruise situation. In either case, the ice-producing layer is often no more than a couple of thousand feet thick (check the forecasts; sometimes there are *many* thousands of feet of icing conditions), so your best bet is to change altitude and change it quickly. When climbing through a relatively shallow layer, watch for the first evidence of icing. When you see it, slow down your rate of climb, let the airspeed build up, then haul back on the wheel and get through it in a hurry. If you encounter icing in a cruise situation, use that turbo power to get to a higher altitude as rapidly as you can.

It's unrealistic to dictate a climb in *every* icing situation, but if you go up, at least you'll have some more altitude to work with if things get worse. One thing completely useless to a pilot is altitude above.

# ✝ Fast Descents

Another payoff of high-altitude flight comes during the descent, when you take advantage of going downhill. A lot of potential energy was stored in the airplane by climbing, so when ATC clears you to descend, stay at cruise power and let the airspeed build. (If your installation doesn't control manifold pressure automatically, reduce the throttle setting as you descend to prevent overboosting.) Descending at cruise power will more than likely push the airspeed indicator up into the yellow caution arc, which is quite all right unless you encounter something greater than light turbulence; if it feels uncomfortable, slow down to a safe speed. (Easy on the throttles, please. Prolonged descents at low power settings are bad news for engines. Keep them warm with a reasonable amount of power.)

Now you must do some calculating; you must adjust the rate of descent to arrive in the terminal area at an altitude compatible with the approach in use. It's easiest to settle on a rate of descent of 1000 feet per minute and let the airspeed stabilize. A groundspeed display is a big help, but in any event, estimate how fast you are moving over the ground versus your 1000-feet-per-minute descent, and you'll have a good idea of how things are going to work out. Cruising at FL 200 and anticipating a rate of 1000 feet per minute, you should request descent perhaps 75 miles from destination; this is based on a groundspeed of about 200 knots, descending to a sea-level airport. If you are at a higher altitude or picking up a real buster of a tailwind, you may want to start down even sooner; ATC is well acquainted with these problems and will usually grant your request.

# ✝ Ear Problems

One thousand feet per minute on the way down is an easily computable number, but sometimes it just won't get you down fast enough. Here's where you must take your passengers' welfare into consideration because, if you descend much more rapidly than that, you can bet that some people on the airplane will leave their ears at altitude, and that can hurt. Fortunately, the rate of pressure increase in a descent is very gradual down to about 15,000 feet, but below this altitude, ear blocks and malfunctioning sinuses can really become a problem. If you find it necessary to limit the rate of descent, let ATC in on the secret. Anytime you or a passenger begins to suffer and the various ear-clearing tricks don't work, level off (advising the controller you have an ear problem on board will get you any altitude you want) and let the pressures equalize. If one of your passengers is *really* suffering, climb a few hundred feet, which will usually relieve the pain, and then start down slowly, making sure your passenger hollers, swallows, yawns, laughs, or does whatever else is required to keep inner and outer ear pressures close together.

Most nonflyers, unaware of the dangers involved, will hide their discomfort behind a shield of pride, and before long, the problem may solve itself, sometimes with a great deal of pain and probably a black public relations eye for you and aviation in general.

Sinuses can be even worse, and you as the house doctor on the airplane should be familiar with the Valsalva and other methods of relief, as well as knowing when *not* to use them (consult a qualified Aviation Medical Examiner). Better yet, don't take yourself or anyone else with colds, hay fever, or other respiratory maladies to high altitudes.

# ✝ High-Speed Approaches

If all has gone well in the descent, there's no need to slow down for the first stages of an instrument approach; in busy terminal areas, where you're being mixed with other high-speed traffic, Approach Control will usually *request* that you keep up your speed as long as possible, sometimes right up to the Final Approach Fix. No problem at all if you're on top of the situation, meaning that you have to shift into "approach gear" very early (see Chap. 14, Getting Ready for an Instrument Approach, for suggestions and good operating practices). You'll most likely be vectored onto the final approach course where things happen in very rapid succession, especially if you are complying with a controller's request to keep your speed up for spacing. Remember that you have been *requested* to maintain a high airspeed, and if you are uncomfortable, uneasy, and just don't think you can handle it, turn down the request. You may be vectored through a less direct approach, but if it's better for you, don't hesitate to leave the high-speed approaches to them that wants 'em.

# 20

# Handling IFR
# Emergencies

In my dictionary, an emergency is "a sudden, unexpected event or
happening; a combination of circumstances calling for swift and decided action"—and I think you'll agree that virtually all of the situations aviators consider "emergencies" fit that description. Probably
more often than not, a pilot has advance notice of an impending
problem, so the "sudden, unexpected" nature of an emergency is not
always present; and while remedial action must be considered and
well thought out, it shouldn't necessarily be *swift*. Moving a control or
switch or lever too rapidly can often make a bad situation worse.

In the IFR environment, a pilot is subject to the same potential
emergencies as in VFR conditions, plus a couple of situations that are
unique to operating in the clouds. The big difference is the *resolution*
of the problem because the IFR airplane must somehow be flown
clear of visibility restrictions before it can be safely brought to Earth.

The redundancy of flight instrument power (DG and attitude indicators are normally driven by air, the turn indicator by electricity, or
vice versa) makes the likelihood of a *complete* instrument failure
rather remote. In the discussion that follows, we assume that the pilot
will always have at least the "partial panel" instruments available to
control the airplane and maneuver it as required.

Yet another assumption is necessary for a sensible discussion of
IFR emergencies; namely, that an instrument pilot encountering an
emergency of any kind while operating in visual conditions will do
whatever is possible and practical to *remain* in VMC until the problem
is resolved. This may result in landing somewhere other than the
planned destination, but remaining VFR and taking care of the problem

sure beat the socks off entering or continuing in IFR conditions and compounding things.

The classification of IFR emergencies is no less difficult than it would be for any other regime of flight because each type of aircraft has its idiosyncrasies, quirks, and unique systems, to say nothing of the very apparent fact that a given situation may be perceived as a life-threatening emergency by one pilot and a mere inconvenience by another; experience and preparation make the difference. However, there are some general groupings of emergency conditions that lend themselves to discussion, to wit:

- Mechanical
- Electrical
- Meteorological
- Powerplant

Before proceeding, be advised that the comments and suggestions offered for the prevention and resolution of these various emergency situations are general in nature; the handling of a specific emergency must always be considered in light of the information provided by the aircraft/accessory manufacturer and the conditions that exist at the time. Doing things "by the book" and in accord with common sense seems a good practice.

# ✛ Mechanical Emergencies

The proper operation of major aircraft systems is at issue here, so if yours is a very simple airplane in that respect, there's not much to go wrong. As speed, power, and overall capabilities improve, there will be more pieces of machinery on board, and that means more potential problems.

Retractable landing gear systems have become less complicated over the years, and the procedures for manual extension (and in some cases, retraction) have followed suit. The instrument pilot who discovers that the wheels will not operate properly is ripe for distraction from flight-control duties, and so the first thing to do in this case is *nothing!* If you've just lifted off into the clouds and the wheels won't come up, leave everything as is and, when the airplane is configured for the climb (power set, trimmed hands-off, autopilot engaged if that's your pleasure), turn your attention to the problem at hand.

Depending on the system, a fail-to-retract condition could result from a blown circuit breaker, faulty microswitches, hydraulic leak, or maladjusted selector switch or lever. Recycling may solve the problem, but what if the mechanism jams in an unlocked condition? Your situation would be sweaty enough in VFR conditions, but why make things worse now that you are in the clouds? In general, if the landing gear fails to retract and the indications show that the wheels are down and locked, *leave them down and locked,* return and land, and take care of the problem on the ground.

During the approach to land at the end of an instrument flight, a failure of the landing gear to extend creates a different kind of emergency. You should do everything within reason to get the wheels down and locked, but *not while you are also trying to execute an IFR approach procedure!* Most emergency gear extension systems require the manipulation of switches, valves, levers, and so forth and most of them require more time than normal, so you'd be wise to consider breaking off the approach and asking for vectors in a wide pattern or an extended holding pattern while you get the gear down.

Refer to the appropriate checklist for emergency gear extension (even in the simple systems); if the hydraulic system permits inflight access to the reservoir, it's not a bad idea to carry a spare can of fluid. Even if it's only enough to operate the system one time, it can save a lot of grief.

Gear won't go down, no matter what? Request clearance to a large airport to take advantage of their emergency facilities, but don't expect a foamed runway because that procedure has largely fallen out of favor. The *Pilot's Operating Handbook* provides the best procedure and configuration for landing with the gear up.

Wing-flap malfunctions are rare but they can be troublesome, especially for those airplanes whose size and speed cause them to depend on flaps to operate onto relatively short runways (no-flap landing distance for light airplanes is usually well within available runway length at most IFR airports).

There is little to be done when wing flaps refuse to extend, and sometimes, continued attempts to operate the system may result in flap extension on one side only, a situation guaranteed to test the mettle of *any* pilot! Know ahead of time what airspeed is required for a no-flap approach and landing; select a longer runway or another airport if you are really concerned about landing distance. Beware of attempting a circling approach with no flaps; you'll be maneuvering at a higher than normal airspeed, tempted to turn a bit tighter, and the higher flaps-up stall speed can creep up behind you and bite.

A third "system" that occasionally provides anxious moments for instrument pilots is the cabin entrance door. When closed and locked, doors are an integral part of the airplane's structure; but when un-latched, they become troublemakers of the first order.

There's perhaps as much psychological disturbance as anything else when a door pops—nearly all airplanes will fly, somehow, with a door open—and the predisposition to get the offending door closed at any cost has resulted in more than one accident. The flat doors found on high-wing airplanes create little more than a noisy distraction when they open in flight; most can be reclosed with little effort, but don't try it *until and unless you're settled down and the air-plane is under complete control.* Even if it can't be closed, there's lit-tle lost; it's noisy, to be sure, but you can return, land, close the door, and continue, this time making absolutely sure the lock is engaged!

Entrance doors on low-wing airplanes are a different story. Nearly all of them are curved to fit the top contour of the fuselage, which means that when partly open the curved upper portion is ex-posed to the slipstream. This may well twist the door and make it nearly impossible to close and latch. But that's not the only problem; the entrance door on a low-wing airplane opens into the area of low pressure directly above the wing and is therefore *pulled* open in flight. In most cases, the unlatched door will stand open several inches, a balance between the positive pressure of forward motion and the negative pressure of the upper wing. With that relatively huge disturbance in the airstream, the tail surfaces are likely to be buffeted severely, with attendant control problems.

The pressure "standoff" that holds the door partially open makes it very difficult to close the door; if you have a strong right-seat pas-senger and care to try, you *may* be able to slow the airplane and get the door closed. But in most cases, it's better to forget the door, fly at a reduced speed (tail buffeting may become a factor), and land as soon as you can.

If a door is going to open, it will most likely do so at takeoff, when aerodynamic pressures begin to work on it. You'll be sur-prised—that's guaranteed—but if there's enough runway ahead to abort or land and stop safely, do so. It's much better to fix the prob-lem on the ground than to fight it in the air.

The failure of a vacuum pump in flight in IMC produces a bona fide emergency situation in single-engine airplanes because it forces you to operate on partial panel until you can reach VFR conditions (see Chap. 4, Partial Panel IFR, for suggestions and techniques). There's one additional consideration here, and that's for the pilots whose single-engine airplanes are equipped with deicing boots; when all the vacuum (or pressure, in some airplanes) is gone, not only are certain instruments inoperative, but wing and tail deicing is no longer available. That's a serious loss in most icing situations, and you must make *immediate* moves to get out of the ice and move even faster than the "immediate" we recommend with boots working.

# ✝ Electrical Emergencies

It may be unrealistic to expect all instrument pilots to be intimately familiar with every system on their airplanes, but those who understand the basics of electrical power production and distribution have a leg up when a malfunction strikes. Electrical power is indeed the lifeblood of IFR operations; with plenty of electricity, navigation and communication are assured, but if you lose electrical power, trouble is guaranteed. Knowledgeable instrument pilots can often spot electrical problems early enough to blunt their effects and are able to manage the electrical system to their advantage when something goes wrong in flight.

Near the top of the electrical emergency list would be a complete loss of *normal* electrical power, which is most likely a malfunction within the alternator or generator itself (a sheared shaft, a broken or loose belt, a breakdown of regulating circuits, and so forth). In any event, all of the electrical systems and accessories will immediately begin drawing from the *emergency* power source, the battery. IFR pilots clearly need to think of the battery as an emergency power source (especially in single-engine airplanes), and one which has very limited capacity. The situation in which the alternator or generator has failed and all the electrical units are drawing from the battery is a situation in which you must take *immediate* steps to shed some of that load and get on the ground ASAP.

Given the small reservoir of electrical power represented by the battery, an alternator or generator failure should be treated as if there were no way to get the dead units back on the line. Shut off everything you don't absolutely need and look for nearby airports with acceptable weather conditions in case you can't restore normal system operation. *Then* go to the *Pilot's Operating Handbook* and meticulously follow the checklist procedure for restarting the alternator or generator; the proper sequence is important to keep from making a bad situation worse.

Most IFR-capable small airplanes are equipped with alternators, and you should know that if there is no voltage in the system (that is, alternator inoperative, battery charge depleted), *you will probably not be able to restart the alternator* (some airplanes have a built-in emergency battery to excite the alternator field and get things going again). It makes sense, therefore, to include a check of the battery's health in your preflight routine for an IFR flight and keep tabs on it throughout the trip.

Once electrical power is restored, it's a good idea to bring units back on line one at a time, checking the loadmeter as each is activated. *Something* caused the alternator/generator to kick off, and it would be very helpful to know if a specific electrical appliance is at the heart of the problem.

Twin-engine airplanes come equipped with more elaborate emergency procedures to cope with electrical power problems; get familiar with the book and use it religiously when the time comes.

Circuit breakers make life easier for the instrument pilot because they readily identify and isolate electrical faults. An open ("popped") circuit breaker represents an electrical emergency only when it is ignored or misused. In the former case, by not including the CB panels in your regular scan you might miss the loss of a critical piece of IFR equipment until the damage is done (this is particularly true with regard to electrically powered flight instruments); in the latter case, the pilot who resets a circuit breaker more than once (or worse, holds it in) is ignoring the warning, bypassing the protection, and asking for big trouble.

In general, reset a popped CB *one time only,* and then only if it's protecting a system or unit essential to the continuation of the flight. *Never* hold a circuit breaker to keep it from popping; that's almost guaranteed to cause problems sooner or later. Some electrical systems are arranged so that certain CBs will blow only when there's a massive short; know which ones (if any) on your airplane fall into this category and should therefore *never* be reset.

When an electrical problem first rears its ugly head during an IFR flight, there's no assurance that it won't go "all the way" and leave you with seriously impaired nav/comm capability. Let the controllers know early on that you have a problem and that, if you "disappear" from their headsets, you are going to continue your flight as cleared, will divert to Airport B, or do whatever the situation demands. Once again, common sense comes to the fore, but your options are more numerous when you've advised ATC of your problem and intentions.

Let's imagine that the electrical gremlins have just about done you in; you're in the clouds, far from destination, can't get out of the weather without an approach to minimums, and operating on battery power alone. Even when you're down to what seems to be the last straw, there's a lot you can do to improve your position and save the day.

First, turn off absolutely everything you don't need to survive, including lights (you *didn't* forget a flashlight with fresh batteries, did you?!), heater, boost pumps, ventilation fan, and *all radios.* Yes, all radios. Of course, let ATC know that you are going off the air, then shut everything down, hold your heading, and turn on one VOR every 5 minutes or so to check your progress; save that precious electricity for the approach, when you'll need one navaid full-time.

Second, remember that transmitting requires considerably more power than receiving, so make only those calls that are absolutely necessary—and then, make them very short. If there's ever a time when clicking the mike for acknowledgment is okay, it's now. The controllers will understand.

Third, don't operate any electrically powered airplane systems in the normal mode if they can be operated manually. *Every* retractable has an emergency gear extension system that uses no electricity, and you can land without wing flaps if need be. At night, save the landing lights until you know that you are close to the runway and then turn them on and hope for the best; any light at all will be helpful at this point.

In that apparently hopeless situation of complete electrical power loss (we're talking *complete* here; no alternator or generator, no battery), you've little choice but to head for VFR weather conditions and hope that you have enough fuel to get there. Says a lot for good preflight planning.

Finally, whatever the situation, don't panic! Remember that the airplane will continue to fly quite nicely without electrical power; it's up to you to stay in control and *keep flying,* right up to the end. These three words—*Fly the Airplane!*—are the most important in *any* aviation emergency.

Earlier in this chapter, we mentioned the difference that experience and preparation can make in the perception of an "emergency" by various pilots. Here's a tale of woe that illustrates what can happen when an electrical *un*emergency is allowed to take charge of a pilot's judgment and actions.

At 300 and 3, the weather was courting ILS minimums, and the forecast was that it would get worse as the Comanche taxied for takeoff. The pilot had logged 750 hours of flying time, including 76 hours of actual instruments and 47 hours in a simulator, and was now embarking on his first IFR trip in the just-purchased

Comanche. His instrument proficiency apparently left something to be desired however because, during an IFR training flight 6 months earlier, the CFII would not endorse an instrument competency check, saying, "...he was rusty on instruments, had trouble intercepting radials, and had difficulty on approaches."

The Comanche had an old horizontal-card DG, which baffled the pilot, according to the airplane salesman, and like many airplanes of that era, the Comanche was equipped with a generator and ammeter. The pilot had experienced a loss of electrical power on the way into Atlanta 2 days before, and the landing gear had to be lowered manually. Power was restored when the pilot fiddled with the circuit breakers; a mechanic found a malfunctioning voltage regulator, fixed it, and signed off the airplane.

Let's take stock at this point; unfamiliar airplane, questionable electrical system, at least one "baffling" primary flight instrument, questionable IFR currency and proficiency, weather going downhill fast, but very good weather within 50 miles or so. Nevertheless, the pilot received his clearance and roared off into the clouds.

Only 12 minutes after departure, he radioed, "We'd like to go back to the airport; we're having a little alternator or generator trouble." He was immediately cleared direct to a nearby VOR in preparation for the ILS approach.

Throughout the vectoring that followed, the Comanche pilot had considerable difficulty in communicating; at one point, he said, "We're all right, we're just getting a drain on our batteries; we'd like direct as possible back to the airport." When he failed to respond to four subsequent heading instructions to turn him onto the localizer course, the airplane had gotten to a point very close to the outer marker and at an angle that made a normal intercept impossible. The controller advised of the proximity and angle and asked if the pilot would like to "come out and try another shot at it?" "We'd like to try to shoot it," said the pilot, and those were his last words. The airplane struck the ground in a steep descent within 1 mile of the outer marker.

Investigation showed that the gyro instruments were operating at the time of impact and that transponder signals were being received and recorded throughout the flight. That information, plus the constant strength and quality of the pilot's transmissions, leads to the conclusion that electrical power was available.

But the pilot had recently experienced trouble with the unfamiliar electrical system, had just gotten the airplane out of the shop, and apparently convinced himself that the problem lay in the generator (or alternator, he wasn't sure which). With a perfectly flyable airplane, why didn't he continue eastbound toward an area of VFR weather instead of turning back into rapidly deteriorating conditions? Why, when advised of a localizer-intercept situation that would tax the skills of a pro, didn't he execute a missed approach with his perfectly flyable airplane and go somewhere else to land?

We don't have the benefit of answers to those questions, but this is a classic example of inexperience, lack of preparation, and *failure to fly the airplane* in a perceived emergency situation. If you *think* you have an emergency, you may have manufactured one.

Lessons to be learned? Know your airplane's systems, don't take a completely strange airplane into IFR conditions, know where *better* weather exists and go there when you need to escape, and above all, when the going gets rough, *fly the airplane!*

# ✝ Meteorological Emergencies

If you use your IFR capability on a regular basis, you will surely encounter some unpleasant weather during your instrument-flying career. Not just low ceilings and visibilities (they're considered routine, and the conditions for which you're trained) but *turbulence* and *structural icing* at a hazardous level. Of course, individual perceptions vary, and no matter how bad you thought it might have been, don't expect to win the "hairiest flying story" contest at the airport; the first liar doesn't stand a chance.

In the real world, it doesn't matter one iota where your Weather Encounter ranks; you were there, you were concerned, and the situation required immediate attention to keep it from getting worse; that's an emergency in anyone's book.

The most likely place to encounter worrisome turbulence is in or near an active thunderstorm, but there are other turbulence producers that deserve just as much attention from the instrument pilot. For example, you should expect turbulence close to the ground whenever surface winds are blowing hard, and it may extend upward for several thousand feet, depending on the strength of the wind, roughness of the terrain, and stability of the air mass. A mountain wave condition is virtually guaranteed to produce tooth-rattling turbulence somewhere in its path. You should also learn to look for large changes in air temperature, which often signal significant wind-shear activity and large amounts of turbulence.

When you have been briefed about turbulence or have figured out for yourself that it's probably out there, prepare yourself and the airplane to handle it. Make sure that there is nothing in the cabin that can move around because it *will*; a thermos bottle or briefcase can become an unguided missile in heavy turbulence. Make sure that you and the other occupants are tied down tight; there's at least one case on record of a turbulence encounter in which a lightplane pilot hit the top of the cabin so hard that a concussion resulted, and he was barely able to get the plane safely onto the ground. Extra-tight seat belts and shoulder harnesses are a *must*.

The proper airspeed is very important; fly too fast, and you risk structural damage (perhaps failure) from gust-induced overloads, while flying too slowly may result in a gust-induced stall; neither is a happy situation on an IFR flight. The *Pilot's Operating Handbook* provides a "Maneuvering Speed" or range of speeds for flying in turbulence, but that's not enough; the airspeed indicator will probably be useless, so you must know the power setting and pitch attitude that will produce the desired speed.

Good pilots always work harder in turbulence to keep the airplane on a smooth, even keel and provide the best possible ride for their passengers; but as the ride gets rougher, control inputs should soften a bit and begin to serve more as dampers, which means a lenient but firm hand on the controls. Don't retrim and don't make big power changes when you get into a sustained up- or downdraft; you need to be ready for a reversal of the condition. If your altitude begins to vary more than several hundred feet, let the controller know so other traffic can be moved if necessary.

When the going gets so rough that you are truly concerned about maintaining control of the aircraft, you must work extra hard to prevent an upset. In addition to the obvious problems created by tumbling the gyros and the *monumental* spatial disorientation that will surely result, you must be prepared to deal with a gross overspeed condition if the airplane is upset and winds up plummeting earthward under power. In a fixed-gear airplane, pull off the power *right away,* and execute your best partial panel critical attitude recovery.

The pilot of a retractable may have an advantage here; the gear can be lowered at the first sign of real turbulence trouble and may provide enough additional drag to keep an upset airplane from building up airspeed at such a rapid rate. Don't be concerned about the maximum gear-lowering speed at a time like this; if you are screaming earthward and nothing else seems to help, throw out the gear and let the wheel-well doors fend for themselves.

When you've got the time (and when your voice has returned!), let ATC know of your plight, and if you feel that a change of altitude or course will be helpful, request—no, *demand*—it. You'll find that controllers have plenty of respect for such requests when the reasons are made known. *Make the reasons known on the first call.* (And don't forget to make a detailed PIREP when it's all over; you don't want anyone else to go through that, do you?)

An icing emergency will probably develop in a more benign manner; a few crystals form on the leading edges, then it's all white out there, and first thing you know, the airspeed has dropped 10 knots. A little more power, then a little more nose-up to maintain altitude, then more power, and when you're staggering through the air on the edge of a stall at full power, you're in deep trouble.

This is the sort of thing that feeds on itself. Each ice crystal creates drag, which requires a higher angle of attack, which exposes more surface to the airflow, which causes more icing, which decreases airspeed, and on and on and on. All the while the propeller blades have been accumulating ice as well, reducing thrust; if the engine air intakes are restricted because of ice, the problem is multiplied. Sooner or later, the airplane may literally fall out of the sky.

"Not to worry," you say, "*my* airplane is fully equipped for flight in known icing conditions. Paid a *bundle* for all that gear." Good investment, but don't count on it to bail you out of a bad icing situation. Somewhere out there is a weather condition that can produce more icing than all that equipment can handle. Exposure is usually the culprit, when low ceilings and cold, wet clouds force you to stay in the icing environment longer than you'd like. Deicing boots do a good job, but after a while, ice may begin to build up *behind* them, forming sharp ridges that can seriously affect lift production. And of course, as airspeed bleeds away and pitch attitude must increase to maintain altitude, ice begins to form on the undersides of wings and fuselage. Drag doesn't care where it roosts.

Through all of this, you should be concerned about letting ATC know of your plight. The controller will surely grant your request for another altitude, but what if you can't transmit because of ice on the antennas? Communication systems are particularly vulnerable to severe icing encounters because ice accumulates much more readily on small-diameter objects such as antennas.

There is a sure way to beat the structural ice monster, but unfortunately, most light airplanes aren't so equipped. We're talking about raw, unadulterated *power* with which you can climb above the icing level, or at least to an altitude where the accretion rate is less. Power to speed you out of the icing, reducing the exposure. Power to keep the angle of attack low and prevent ice from forming on unprotected airplane surfaces. If you have the power, *use it* when serious icing threatens; know where the most likely ice-free altitudes are and go there ASAP.

No deicing equipment, no surplus of power? You're not going to like this, but you shouldn't be there in the first place in such an airplane. In case you *do* wind up in the icebox, there are still some things you can do to make the most of a bad situation:

1. Let ATC know right away that you are encountering icing and are not equipped to handle it. Confession is not only good for the soul; it lets the controllers begin planning ahead to accommodate you.

2. Request (demand, if necessary) a new altitude when you first sense trouble. It's usually better to climb, but that's not a panacea; your preflight briefing should have alerted you to temperatures aloft, cloud tops, and so forth to help you make a rational decision.

3. Maintain airspeed as high as practical in level flight to reduce underbody and underwing icing. If you decide to climb, do so with dispatch. If you decide to descend, pull out the stopper; you've nothing to lose.

4. Watch engine performance closely. Iced air intakes may shut down small engines and will trigger the automatic alternate air systems on larger ones; either way, a power loss is inevitable. If carburetor heat is required to keep the engine running, there's *another* power thief.

5. Change power (or rpm) frequently, rapidly, and through a wide range in order to flex the prop blades and shed most of the ice they've collected. The props are your lifeline; treat them well.

6. Choose the nearest suitable airport (in terms of facilities and weather) and request vectors to go there *now*. The word "icing" will get the full attention of any controller in the business; say that you're falling out of the sky, and you can have anything you need.

7. Plan ahead so that you can execute the approach procedure in the shortest possible time. Request vectors to the final approach

course to save time and advise each successive controller of your problem. Don't forget that when you break out of the clouds you may not be able to see because the windshield will probably be covered with ice. That's what side windows and credit cards are for.

8. Don't maneuver any more than absolutely necessary once the runway is in sight and *don't extend flaps for landing.* An airplane coated with ice is an experimental vehicle, ready to stall/roll/sink/snap; you're working with an unknown quantity at this point. If it flies without flaps, fly it onto the runway and be happy you're home in one piece.

# ✝ Powerplant Emergencies

Do you recall your most recent airplane engine failure? That question can be answered emphatically by some (that unfortunate, precious few who have actually suffered such a calamity), but for nearly all of us, the reliability of today's powerplants is very much taken for granted. Otherwise, we'd not be willing to risk all on an IFR flight in a single-engine airplane.

The odds improve somewhat with a twin, if for no other reason than the second engine provides a spare source of electrical power for the nav/comm systems to complete the flight. (There will always be pilots who rail against the light twin as an unnecessarily complicated machine with a horrible engine-out safety record, but training and preparation can answer the critics.)

In any event, two cases present themselves when considering IFR powerplant emergencies: *complete* engine failure and *partial* loss of power. Single- and multiengine procedures differ significantly, so let's take them one at a time.

Instrument pilots in the clouds in a single-engine airplane have hung their hats on the continued operation of that powerplant; it not only generates the thrust to keep the plane airborne, but it supplies power for all of the navigation, communication, and flight instruments. A complete engine failure requires decisive, proper action because the options are so limited.

When the only thrust-producer on board quits, your first concern *must* be to maintain airspeed. If you haven't burned the best glide speed number deep into your brain, do so right now because it's a lifesaver. Each knot either side of the proper speed does nothing but increase the sink rate, and that will cost you precious time and altitude. Most singles, especially the high-performance models, cruise at airspeeds well above best glide speed, so if you have the presence of mind, pull up the nose and *zoom*; trade some of that airspeed for altitude. You'll be grateful for every extra foot when you get to the bottom of the inevitable descent.

Airplanes with constant-speed props can be shifted into "overdrive" by moving the prop control to the low rpm position; in this condition, drag is reduced significantly, and glide distance (or time in the air, if you choose) is increased. Time is important for developing and implementing options on the way down.

Declare an emergency as soon as you're squared away so that the controllers can send the St. Bernards to the right spot, to say nothing of their ability to point out nearby airports, provide assistance with vectors, and so forth. In most such cases, you're very much on your own because of the severely limited performance of your airplane-turned-glider, but give the controllers a chance to help.

Preservation of battery power is academic at this point. You'll probably be on the ground—one way or another—long before the battery plays out. But if your ELT is wired into the airplane system, you might give some thought to shutting down electrical appliances to save as much energy as possible, especially if you have invested in an ELT with voice-transmission capabilities.

Where to glide will have to be determined on almost a minute-by-minute basis because the "nearest airport" changes constantly as you move along. It's always best to head for an airport; you *might* make it and, if nothing else, you'll be flying in a straight line, making search and rescue that much easier. It certainly makes sense to avoid high ground, which underscores the value of knowing the terrain over which you're flying.

Unless the weather beneath is absolutely zero-zero, you'll break out of the clouds sooner or later, and the situation has instantly reverted to a plain vanilla forced landing exercise. You haven't practiced emergency landings for years? Take time to retune that skill because a single-engine pilot can *never* be too good at forced landings. If the weather underneath is so bad that you must accept whatever landing site appears at the last minute, you might consider not flying IFR in such conditions. What if you're looking at downtown Pittsburgh when you break out of the clouds?

A *partial* power loss creates a completely different set of circumstances for the single-engine instrument pilot. Now there's at least the hope of continuing along the way, limping to be sure, but not committed to a forced landing. You should occasionally practice flying under the hood at drastically reduced power settings to simulate a partial power situation; you may be pleasantly surprised to find that you can remain aloft at ridiculously low airspeeds. Even if you have to settle for a slow descent, you've bought some time instead of the farm.

Make ATC aware of your plight and then begin to troubleshoot. Each airplane, each engine, each atmospheric condition is a bit different, and you may have to run through a lengthy litany of checks on boost pumps, fuel selectors, carb heat, throttle settings, whatever. Keep in mind that an airplane engine seldom fails without warning; the smart pilot knows the importance of engine-instrument indications and includes them in a regular, frequent scan of the panel.

Multiengine pilots certainly have the powerplant-failure odds working for them, but the price for all that capability is the additional training and skill required to keep the machine flying properly with one engine shut down. In addition to a primary concern for the proper airspeed (not less than Vyse), multiengine pilots must maintain heading, get the offending engine secured, and make plans for an approach at a nearby airport. Whenever you suffer the loss of an engine—VFR or IFR—consider yourself in an emergency situation and *get the airplane on the ground as soon as you can.* Trying to stretch the flight to the home base (where repairs are less expensive) or the original destination (where the passengers think they need to be) has brought multiengine pilots to grief more than once.

Safe engine-out IFR operations consist primarily of proper management of the airplane and its systems. You'll need to maintain a zero-sideslip attitude to get optimum performance, and that will require trimming in all axes; your attention will wander from constant control pressures, and you'll wind up fighting the airplane. Let the trim do its work.

Shut down nonessential electrical units to reduce the load on the operating alternator/generator (check the POH for advice and guidance). Fuel management should not be a problem unless the engine failure has taken place in a remote area where you'll need some of the fuel on the dead-engine side to get out of trouble. Study and practice the crossfeed procedure well ahead of time, and *use a checklist* when you begin changing selectors and crossfeed valves; it will get mighty quiet if you make the wrong moves.

An engine-out instrument approach should be flown just like a normal one, at least in terms of airspeeds and IFR procedure. Of course, you'll want to use every opportunity to cut out excessive maneuvering, and hold up on gear extension until you are sure you can land. A single-engine go-around is hazardous at best, so choose a destination with landable weather. Don't compound your problem.

With an engine out, airspeed is the key to your survival; *never less than Vyse until landing is assured* is a good rule of thumb. In cruise, you may have to settle for a slow rate of descent at Vyse, but it's likely that at some lower altitude the wings will develop enough lift to maintain level flight, so perhaps now you can reduce power a bit and save the remaining engine.

Following an unsuccessful attempt to complete the approach at his planned destination in northern Florida, the pilot of a chartered Beech E-18 elected to deliver his passengers to Gainesville. That was a good choice; with a ceiling of 400 feet and a mile and a half visibility in fog, a successful approach was all but guaranteed, and the passengers could rent a car and still keep their schedule. There was an atmosphere of "hurry-up" as the pilot approached the outer compass locator from the northwest at a high rate of speed, cleared for the ILS for Runway 28.

Rather than proceed outbound far enough to set up a stabilized approach, this pilot—a highly experienced exmilitary aviator—elected to shortcut the procedure with a modified 90-270 as soon as he crossed the marker. Predictably, the high-speed turn carried the airplane well beyond the localizer course. A descent was begun on the inbound heading, even though the radar plots showed that the airplane was never established on the localizer.

Leveling off several hundred feet above Decision Height, the pilot continued past the airport, well north of the localizer, out of sight of the runway, and still in the clouds; he reported "outer marker inbound" when the airplane was actually passing the middle marker.

And then, when the pilot decided to execute the missed approach (his best decision of the day), the trap closed. Within seconds of the "missed approach" call, Tower controllers heard, "I've lost an engine." After several unanswered calls from the Tower, the pilot indicated that he was still there, but that his troubles were far from over; he couldn't get the prop feathered.

A Beech E-18 (that's one of the old-timers, with two round engines up front and the little wheel in the back) is no sky-rocket on one engine in the best of circumstances; toss in a windmilling prop and a full load and this airplane isn't likely to make *any* vertical moves except down.

The postaccident radar plot showed the airplane continuing westbound at an altitude of not more than 600 to 700 feet above the ground (still in the clouds, of course), and finally, some 7 miles west of the airport, the pilot apparently lost control; he probably encountered $V_{mc}$ and $V_{stall}$ at the same time. The airplane entered a rolling, diving turn to the right and was ripped apart by the guy wires of a large television tower. No survivors.

This was clearly a powerplant emergency made almost irrecoverable by a propeller system that failed completely when it was needed most. The pilot seemed a victim of fate, but careful investigation turned up some intriguing facts. A filter element was completely missing from the right engine fuel system and was perhaps the initial event in the engine-failure sequence. When the right engine was opened up, an uncommon amount of sludge and dirt was found in the oil screens...the same oil system that provides pressure for prop feathering. There had been a number of other squawks on this airplane in recent weeks regarding various mechanical problems, squawks that had not been corrected.

Finally, the autopilot was known to be inoperative; not that it would have saved the day, but this was a Part 135 operation which required either a functioning autopilot or a flesh-and-blood copilot. Could a little bit of help in that harried cockpit have made the difference when things came right down to the wire?

The moral of this story is simple and direct: When you *know* that the airplane you're about to take into instrument conditions is in less than top shape *in all respects,* you (and your passengers, through you) are assuming a terrible risk. There's always a better way; get a substitute aircraft, wait until the weather gets decent enough to accommodate the risk, or insist that the airplane be repaired before you go.

And because this pilot's original destination was illegal under FAR Part 135 (no weather-reporting service, weather below minimums), it wouldn't have hurt a bit to go by the rules—and not go at all.

# 21

# Proficiency Exercises

A lot of rust and bad habits can form in the 2 years between Biennial Flight Reviews, and often the only time general aviation pilots find out that they are not quite as sharp as they ought to be is when they get into a bind. In most cases nobody else knows about it, but on occasion the headlines proclaim the lack of proficiency. When you get right down to cases, is your neck less valuable in your own airplane than when you are a passenger on an airliner? You place complete faith in the pilots with four stripes on their sleeves because you know they must maintain a high level of competence; there's no reason why you should not demand the same of yourself when *your* hands are on the controls.

Flying, particularly instrument flying, is a sophisticated skill, and unless you fly IFR regularly and frequently, you *know* that you must practice that skill to be as good as you want to be. An occasional IFR cross-country won't do the job because a typical "real-world" flight consists of following radar vectors and airways until you get into the terminal area, then some more radar vectors to the final approach course, and relatively simple navigation to the runway. When the chips are down and you are required to fly the approach all by yourself with procedure turns, climbs, and descents to predetermined altitudes and headings, missed approaches, and no-radar navigation, your rusty techniques don't do you much good.

There's only one way to be sure, to be safe, to be sharp, and that's a regular program of practice that will lead to precise control of the airplane almost as an afterthought, with most of your attention and thought processes devoted to staying ahead of the situation. If you're a beginner, the exercises in this chapter will help *get* you that way, and if you're an experienced IFR pilot, they will help *keep* you that way. Whether you need to use all or a selected portion of these maneuvers will depend on your particular needs, your skill level, and how often you fly on the gauges. Are you the type who likes to accept a challenge? Run through the entire series next time you're under the hood; you'll soon know whether or not you need the practice.

# ✝ Exercise #1: Straight and Level Flight

It's more difficult to keep the flight instruments standing still than it is to make them move in the proper direction at the proper rate. Flying straight and level, with no changes in heading or altitude or airspeed, is a great deal harder than turning, climbing, or descending, but it's not impossible, so set yourself up in level flight and adjust the little airplane on the attitude indicator so that it rests exactly on the horizon; that's your best reference for what's happening or for what's about to happen. As long as you keep the pitch attitude where it belongs and the wings level (the ball in the turn-and-bank is assumed to be centered all the time), you will maintain straight and level flight. The altimeter will indicate immediately any climb or descent, and the directional gyro will keep you honest in the heading department. The very instant you detect any change on either of these instruments, apply corrective pressure to stop the movement and then apply additional pressure to return things to their proper places.

The procedure is very simple and should become an unconscious, three-step method to be used in all your instrument flying. First, cross-check the instruments (all of them, as rapidly as possible but with emphasis on the attitude indicator); second, interpret the indications so that you can decide what to do about it; third, apply control pressures to control the situation.

Straight and level flight is difficult because your task is to keep things from happening; it's satisfying because it will develop a mental discipline that will carry through to other maneuvers. If you can do a good job of straight and level flight, you're well on the way to precision instrument flying. Stay at it until you can keep the needles steady, narrowing your tolerance as you progress.

# ✛ Exercise #2: Standard-Rate Turns

Starting from straight and level flight on a cardinal heading, practice turning at the standard rate of 3 degrees per second. The secret of smooth, accurate turns is *pressure*; all that you should have on your mind is applying enough aileron pressure to make the miniature airplane begin to bank. A smooth, slow increase in bank attitude is what you're looking for, and as long as you maintain the pressure, the airplane will continue to bank, so keep pressing until the turn needle indicates a standard-rate turn.

You should know the angle of bank required for a standard-rate turn at two airspeeds—cruise and approach. There's an easy rule of thumb that you can apply: Divide the *true* airspeed in knots by 10 and add 5. This number is the approximate bank angle for a 3-degrees per second turn. For example, cruising at 130 knots true airspeed, bank 18 degrees for a standard-rate turn.

At this point, relax the bank pressure, notice the amount of bank on the attitude indicator, and make small corrections as necessary to keep it right there. As instrument indications begin to change, cross-checking becomes vitally important; banking the airplane changes the distribution of lift produced by the wings, so as soon as the altimeter moves, apply back pressure to stop it.

The heading indicator should be consulted during the turn, but only with a glance, since your objective is to maintain the bank angle that will produce a standard-rate turn. Don't forget about the heading completely, because you want to roll out on a certain number. You may want to experiment with this a bit; begin the roll-out when the heading indicator passes through a number that is one-half of the number of degrees of bank from the desired heading. If your standard rate requires 20 degrees of bank, start your roll-out pressure when you are 10 degrees from the heading you want. Use the same, slow, deliberate process that got you into the turn, remembering to remove any back pressure you needed to hold altitude; as the wings return to level, the lift you redistributed in the turn will show up as a tendency to climb. A rapid, complete cross-check will help you stop the altimeter before it moves very far.

As soon as the turn needle lines up with its center index, you've stopped turning, and if you've done it properly, you will run out of bank at the same time the heading indicator comes to rest on the sought-for number. Remember that you are maintaining constant pressure to reduce the bank attitude, which produces a constantly reducing rate of turn. After a few practice turns, you should know just how much lead to use to make everything come out even.

Most pilots have a strong tendency to roll out of a turn considerably faster than they roll in, so you may have to force yourself to slow down. Slow, smooth application of pressure to effect the desired change in attitude is the secret. You'll soon be making turns that will not even be noticed by your passengers when there are no outside references; this is the target you should set for yourself.

## ⟊ Exercise #3: Steep Turns

A safe limit for most pilots in actual IFR conditions is 30 degrees of bank, but there's nothing like the ability to handle your airplane in a steep turn to build your confidence. Start your practice with enough bank to push the turn needle a little past the standard-rate mark, and as your skill builds, keep it going until you can fly around confidently and smoothly in a 45-degree bank, rolling out right on the headings you desire.

There are a couple of things to watch for. Roll into the turn at a slow, deliberate rate, and as the bank angle progresses, pay more attention to the altimeter; you are changing the lift situation rapidly as the wings get further and further from the horizontal. A steep turn requires a great deal of back pressure, and it's not a bad idea to roll in some trim to help you maintain altitude. Your lead on the roll-out headings will have to increase as well to compensate for the faster rate of turn. When roll-out time comes, use the same slow, deliberate pressure that got you into the turn. And don't forget about all that back pressure and elevator trim.

## ⟊ Exercise #4: Constant Airspeed Climbs and Descents

Before you can start this one, you'll have to go back to the drawing board to discover the attitude that will produce the best rate-of-climb speed. When you have it nailed down, make a mental note of the picture you see on the attitude indicator; within the limits of available power, you can rest assured that when you apply climb power and rotate the little airplane to that predetermined position, the same airspeed will result.

Next, set up an attitude and power setting for practice descents. Because this is an exercise, not an approach situation, leave gear and flaps up, reduce power to just above idle, and experiment with the pitch attitude until you are descending at 500 to 1000 feet per minute at the same airspeed used for climbing. Once you've established this, remember the attitude and power setting.

Now you're ready to go to work. From straight and level flight at normal cruise, simultaneously increase the attitude to the predetermined position and smoothly add power to the climb setting. Don't worry about airspeed, it will take care of itself; but you *will* have to increase the speed of your cross-check to keep the heading indicator from wandering off to the left. Feed in whatever rudder pressure is required to maintain your original heading. Sail right on up for 1000 feet; don't change a thing until you are within 50 to 100 feet of the desired altitude and then *smoothly* press the wheel forward until the little airplane rests on the horizon bar and hold it there, using rudder pressure as needed to control heading. As the airspeed builds up to cruise once again, reduce power to normal and that's all there is to it!

The higher the climb performance, the more you will have to lead the level-off. Lead the altitude by 10 percent of the climb rate— 50 feet when climbing at 500 feet per minute, 100 feet when the climb rate is 1000 fpm, and so on. When the altimeter comes to rest on the desired altitude, it becomes the primary source of pitch information during the time it takes to regain cruise airspeed.

Descent technique is only slightly different, but of course, everything works in reverse. From normal cruise, reduce power to the descent setting you worked out earlier and hold the little airplane on the horizon bar until the airspeed approaches the proper number for descent. You'll notice a need for left rudder pressure to keep the heading constant, and as your cross-check shows the gyro numbers creeping off the mark, press with the appropriate foot. When the airplane is slowed to the speed you want, *smoothly* press the nose down to the attitude you set up in your test, and you're on your way.

Descend 1000 feet, but before you get to the bottom, consider the effect of both gravity and inertia resisting your attempts to level off; increase your lead by 50 feet, at which time you should begin a slow addition of power and a *smooth* increase in pitch attitude to bring the little airplane to rest on the horizon. As in the climb, you'll have to experiment a bit to find the proper lead for your airplane, the airspeed you're working with, and your personal technique. Coordination and *smooth pressure* are the keys.

# ✝ Exercise #5: Constant Rate Descents

Descending at a constant, controlled rate is useful when you must reach an MDA within a certain distance from the Final Approach Fix. The objective is to set up the descent airspeed and then control the *rate* of altitude change with power. A reasonable figure for most light aircraft on most approaches is 500 feet per minute, but a particular situation may call for a higher or lower rate of descent. Make the power changes small, keep the little airplane steady in the descent attitude, and remember how much manifold pressure or how many rpms are required. Go down 1000 feet and accomplish a normal level-off.

You may have noticed that 1 inch of manifold pressure or a 100-rpm change produced roughly a 100-feet-per-minute rate change in both exercises. If the power change turns out to be more or less for your airplane, remember the figure; it will smooth out your glide slope corrections later. This is also a good time to calibrate the vertical-speed indicator by checking the altimeter readings at the beginning and end of a fixed time period, and you'll know how much your VSI is out of calibration, if at all.

# ✝ Exercise #6: The Vertical S

This is a coordination builder that requires constantly changing thought patterns and forces you to plan ahead. Like flying into a funnel without touching the sides, the Vertical S gets more demanding as you proceed. It starts with a 500-foot altitude change, next time it's only 400 feet, then down to 300, and finally 200 (Fig. 21-1). You'll hardly have time to make the attitude and power changes for a 100-foot dip, so be satisfied when you can go through four successive ups and downs.

Start from normal cruise and descend 500 feet at constant airspeed. When level-off time rolls around, *don't!* Keep the nose coming right on up to the climb attitude while you're adding climb power. The objective is to touch the bottom altitude and enter a climb without any change in airspeed. Success in this exercise is a matter of dis-

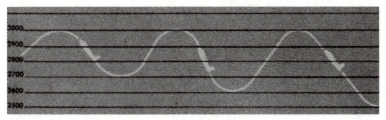

**Fig. 21-1.** *The Vertical S.*

covering the right combination of altitude lead, rate of pitch and power change, and a soft touch on the rudders. Now climb back to the original altitude, *start* the level-off at the normal lead point, but *don't level off*; instead, press right over into a constant airspeed descent. You're trying to brush the top altitude before starting down again. From here on, it's simply a repetition of what has gone before, but each time you bottom out, chop off 100 feet.

## ✝ Exercise #7: The Vertical S-1

If you happen to be a roller-coaster nut and dream of the big day when you get to ride the world's wildest—one that not only goes up and down, but changes its direction of turn every time—dream no more! Your wishes have come true in the Vertical S-1, and the only disappointment will be that you won't get to see what's going on; you'll be inside, under the hood, making it happen. This is the last "challenge" maneuver in the series, and when you master it, pat yourself on the back; you are doing a good job of flying your airplane on the gauges.

Start from normal cruise, but don't begin any maneuvering until you reduce airspeed to the climb/descent speed used in Exercise #4. When this is achieved, take a deep breath, and enter a descending standard-rate turn. So far, much like Exercise #6, but the similarity ends when you have descended 500 feet; at this point, repeat the Vertical S "soft bounce," enter a constant airspeed climb, and *at the same time, reverse the direction of the turn*. In other words, if you started the S-1 with a descending right turn, you should go down 500 feet, turning all the time at a standard rate, and then climb 500 feet while turning to the left. When you reach the top, enter another descent and reverse the direction of turn again. Continue through the same series of altitude changes as in the Vertical S, changing the turn each time you transition from climb to descent or descent to climb.

Don't be concerned about how many degrees you turn during the exercise; the objective is to accomplish smooth, positive, and coordinated changes in aircraft attitude at the proper times.

# ✝ Exercise #8: Simulated Holding Pattern

Now you can begin applying your instrument artistry to more practical matters, such as a standard holding pattern. Get squared away on a heading in straight and level flight and, at a cardinal point on the clock, simulate station passage. Using the four Ts (*Time, Turn, Throttle, Talk*), enter the holding pattern by noting the *time,* roll into a standard-rate *turn* to the right, *throttle* back to holding power, and pretend you are *talking* to ATC, making the required report entering a holding pattern. This procedure will form a habit that will stand you in good stead later on.

The first turn when entering the racetrack is more than just a standard-rate turn because you are also changing airspeed and are required to hold altitude precisely. You will notice that all three controls call for changes in pressure. You need a higher angle of attack because of the lift change in the turn as well as the decrease in power and airspeed; unless you decrease the bank angle, the rate of turn will increase as airspeed decreases, and rudder pressure will be needed as you change power and bank. Roll out at the 180-degree point and fly level for exactly 1 minute. Now another 180 degrees to the right, fly for 1 minute, and you're finished.

You could go on flying in the racetrack pattern all day long, but it wouldn't prove much; you should be aiming for smooth entry technique. Try a couple of left-hand patterns for kicks.

# ✝ Exercise #9: The 45-Degree Procedure Turn

When a procedure turn is required, this one comes as close to standard as anything else. From straight and level flight at your approach airspeed (clean airplane), roll into a standard-rate 45-degree turn to the right and fly for 1 minute. When that time has elapsed, turn left 180 degrees; the second turn will *always* be made in the opposite direction, taking you away from the station and helping to guarantee that wind will not blow you back toward the airport shortening the time available to line up on the final approach course.

At the completion of the 180-degree turn, maintain heading for 30 seconds and then turn left 45 degrees to put you back on course, but going in the opposite direction. That's the entire purpose of the procedure turn: to get you turned around on a specific course. Headwind or tailwind will change the time on the 30-second leg when you are actually intercepting an approach course, but at least this exercise will instill the principle.

For variety, try a couple of procedure turns to the other side of the course, always making the second turn in the opposite direction.

# ✝ Exercise #10: The 90- 270 Procedure Turn

When time is of the essence, this little gem will get you turned around quickly, and it's a good exercise in coordination. To get the most out of it, plan to make your turns greater than standard rate; use 30 degrees of bank throughout. At approach speed in straight and level flight, roll into a turn to the right and begin the roll-out pressure as if you were going to stop the turn at the 90-degree point.

But instead of leveling the wings, keep right on going into a 30-degree bank in the opposite direction. Continue the turn through 270 degrees and bring the airplane back to straight and level on the reciprocal of the original heading.

So you won't develop a right-hand pattern "groove," do a few of these to the left as well. Once again, the second turn is *always* made in the opposite direction.

# ✝ Exercise #11: Modified Pattern B

Gray-haired military pilots will remember that the Pattern B (Fig. 21-2) was a sort of "final exam" of instrument technique before graduating to approaches and it will serve that purpose well for you. In addition to including all of the maneuvers you have been through in the previous exercises, this one requires a good bit of planning ahead, the essence of good IFR operations.

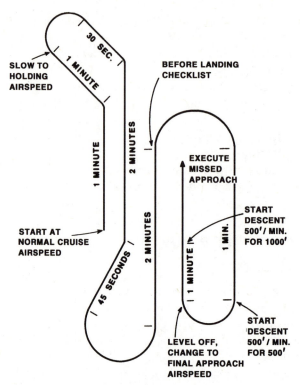

**Fig. 21-2.** *Modified Pattern B.*

Because of its complexity, sketch out the pattern and its instructions on a card that you can clip to the control wheel, to your knee board, or wherever you plan to put approach charts when you get to the real thing.

There is one ground rule for Pattern B that makes it a little different from the other exercises. All timing is done with reference to cardinal times on the clock, regardless of the bank attitude of the airplane. If you find yourself getting increasingly more behind, check your rate of turn and the rate at which you are rolling into and out of turns. The whole pattern is based on precise 3-degree-per-second turns; no more, no less. Start the pattern on a north heading; this exercise is *not* a test of your ability to add and subtract!

At the end of the final leg, you are pretending that you've just arrived at the missed approach point, and unless the hood falls off, you'll still be on instruments, so execute a missed approach. The objective is to accomplish the transition from descent to climb as smoothly and rapidly as possible, with a minimum loss of altitude. Climb out for 500 feet and take off the hood; you've been hard at it for almost 15 minutes of concentrated IFR practice, and you deserve a break.

# ✝ Exercise #12: Unusual Attitudes and Stalls

There is a purpose to be served by practicing recovery from unusual attitudes, but the purpose is *not* the satisfaction of an instructor's sadistic bent. Ever since instrument flying instruction was invented, there have been CFIIs who take delight in complicated, stomach-turning, vertigo-producing attitudes that rank as truly unusual, but not necessarily useful in helping the student become a safer pilot. It is doubtful that in the course of an actual IFR flight you will allow your attention to wander completely away from the array of gauges on which your very life depends. So why should you practice recoveries from attitudes that are suddenly thrust upon you after a period of sightlessness during which the instructor has been flying the airplane?

Since complete inattention is impossible to simulate as long as you can see, it is expedient for you to close your eyes while the unusual attitude is set up, but *you* should fly the airplane into the abnormal situation. You will have sensations of movement but no idea of how much, and after a couple of turns, you likely will not be able to tell which way.

Under the direction of the instructor or the safety pilot, enter and roll out of turns in each direction; you may think you're doing just great, but sooner or later your inner ear and deep muscle senses will begin to fool you, and when your companion recognizes an attitude other than one required for normal instrument maneuvering, it's recovery time. You may have gotten yourself into a nose-low, 60-degree bank or a wings-level, nose-high attitude. The airplane will be headed for the ground and about to go through the redline, or it could be valiantly struggling upward, ready to stall. The important part of the exercise is that you have flown yourself into the situation, and now it's up to you to get yourself out of it.

These two attitudes—nose-low with airspeed building and nose-high with airspeed decreasing—are the themes upon which recovery procedures are based. Allow the former to continue building and the airplane may come apart if it doesn't collide with the ground first; ignore the latter and the airplane will stall.

This is *not* the place for "do anything, even if it's wrong." A strong pull on the wheel when you recognize the high airspeed may be followed by a sharp report as the wing tips meet directly above the cabin. With this in mind, force yourself to make a thorough but rapid check of the instrument indications before you take action. There are only two basic problems, and you can experience only one at a time, so things aren't really so bad after all. And even more to your advantage, the same control is the one to reach for first in either situation; if the airspeed is high and moving higher, reduce power immediately, rapidly, and significantly. Looking at the instruments and recognizing a near stall, your hand should instinctively go to the throttle and keep right on moving, adding power to get you out of trouble.

If the wings are not level when "recover!" sounds (and they probably won't be), apply aileron pressure to get things back where they belong. The only problem yet to solve is that of pitch, which in either case should be returned to the level flight attitude, but the sequence is important, especially in a nose-low, high-airspeed situation. Unload the wings by taking out the bank before easing the nose up to the horizon; it doesn't matter quite so much in the nose-high problem, since the addition of power will probably keep you from stalling.

The ideal recovery from any unusual attitude consists of a quick and accurate interpretation of instrument indications followed by a coordinated, rapid return to level flight with little loss of altitude, direction, or aplomb. It seldom works out that way, especially the problem of which controls take precedence in a particular situation. For the sake of maintaining structural integrity, organizing your thoughts, and providing guidelines, here's the order in which you should get things done when you recognize an unusual attitude (some bank will usually be present):

## First Case: Nose Low, Airspeed Increasing, Steep Bank

1. Reduce power.
2. Level the wings.
3. Pitch attitude back to level (smooth and easy does it).

## Second Case: Nose High, Airspeed Decreasing, Steep Bank

1. Add power (full power, if necessary).
2. Pitch attitude back to level (smooth and easy again, but it's not as critical here as in the high-speed case).
3. Level the wings.

While you're doing unusual things, run through a few stalls while under the hood. Stalls? Under the hood? Why not? You know that you have plenty of control in a stall when you can see outside, and there's no difference with the blinders on. Set the power, gradually bring the nose up, keep the wings level with aileron, nail the heading indicator on the proper number with rudder pressure, and when the airplane shudders and quits flying, make a normal recovery; it's just another unusual attitude. Try stalls in various aircraft configurations and power settings and with different amounts of bank.

You'll be a lot more confident when you can handle all these situations, and it's nice to know that if you ever encounter a stall in actual IFR conditions you are prepared to solve the problem.

# Index

# About the Author

Richard L. Taylor (Dublin, Ohio) is an award-winning author of many aviation articles and books. Currently an aviation consultant, he taught at all levels of the aviation curriculum as an assistant professor of aviation at Ohio State University.